Forensic Botany

Forensic Botany

A Practical Guide

David W. Hall, Ph.D.
David W. Hall Consultant, Inc., USA

Jason H. Byrd, Ph.D.
*William R. Maples Center for Forensic Medicine,
University of Florida, USA*

**WILEY-BLACKWELL**

A John Wiley & Sons, Ltd., Publication

Library of Congress Cataloging-in-Publication Data

Hall, David W. (David Walter), 1940-
    Forensic botany : a practical guide / David W. Hall and Jason Byrd.
        p. cm.
    Includes bibliographical references and index.
    ISBN 978-0-470-66409-4 (cloth) – ISBN 978-0-470-66123-9 (paper)
1.   Forensic botany–Handbooks, manuals, etc.   I. Byrd, Jason H.   II. Title.
    QK46.5.F67H25 2012
    363.25'62–dc23

                                                                        2011048941

A catalogue record for this book is available from the British Library.

Wiley also publishes its books in a variety of electronic formats. Some content that appears in print may not be available in electronic books.

Set in 10.5/12.5pt Times Roman by Thomson Digital, Noida, India
Printed and bound in Singapore by Markono Print Media Pte Ltd

First Impression 2012

Contents

Contributors

Dr David W. Hall
David W. Hall Consultant Inc.
3666 NW. 13th Place
Gainesville, FL 32605
USA
Tolkos@aol.com

Dr Jason H. Byrd
Associate Director
William R. Maples Center for Forensic Medicine
College of Medicine
University of Florida
PO Box 147100
FL 32610-0275
USA
jhbyrd@forensic-entomology.com

Dr Matthew A. Gitzendanner
University of Florida
Department of Biology
PO Box 118526
Gainesville, FL 32611-8526
USA
magitz@ufl.edu

Dr Christopher R. Hardy
Director of the James Parks Herbarium
Millersville University
Millersville, PA 17551
USA
Christopher.Hardy@millersville.edu

Dr Richard Old
XID Services, Inc.
PO Box 272
Pullman, WA. 99163
USA
rold@pullman.com

Bernard A. Raum, JD, MFS
University of Florida
PO Box 1149
Newberry
Florida 32669
baraumlaw@cox.net

Dr Anna Sandiford
Forensic Science & Research Ltd
PO Box 17-317
Greenlane Auckland 1546
NEW ZEALAND
sandiford@theforensicgroup.co.nz

Dr William Stern
Department of Biological Sciences
Florida International University
3000 NE 151st Street, MSB 357
North Miami, FL 33181
USA
landb0801@gmail.com

Dr John R. Wallace
Department of Biology
Millersville University
PO Box 1002
1 South George Street
Millersville, PA 17551
USA
john.wallace@millersville.edu

Series Foreword

Essentials of forensic science

The world of forensic science is changing at a very fast pace. This is in terms of the provision of forensic science services, the development of technologies and knowledge and the interpretation of analytical and other data as it is applied within forensic practice. Practising forensic scientists are constantly striving to deliver the very best for the judicial process and as such need a reliable and robust knowledge base within their diverse disciplines. It is hoped that this book series will provide a resource by which such knowledge can be underpinned for both students and practitioners of forensic science alike.

The Forensic Science Society is the professional body for forensic practitioners in the United Kingdom. The Society was founded in 1959 and gained professional body status in 2006. The Society is committed to the development of the forensic sciences in all of its many facets, and in particular to the delivery of highly professional and worthwhile publications within these disciplines through ventures such as this book series.

Dr Niamh Nic Daéid
Reader in Forensic Science, University of Strathlcyde,
Glasgow, Scotland, UK
UK Series Editor

Prologue: the begining

How does one become a forensic botanist? In the late 1960s and early 1970s the senior author of this book was employed as a plant identification specialist by the University of Florida, eventually becoming the Director of the Plant Identification and Information Services. To be able to help with a plant problem, identification of the plant is an important first step. A plant's name is the key that opens the door to all the known information about any particular plant. Is it poisonous to humans or livestock? Can it be eaten? How big will it grow? Where should it be planted? Is it a weed? How fast will it grow? Is it native? After identification the plant samples were often sent to other experts to determine the disease affecting it or the insects eating it. If the plant was cultivated, horticultural experts were frequently needed to determine the type of hybrid and the care needed for best growth. Pest control, including diseases and/or insects, for cultivated ornamental plants was a constant request. Even more vital is the control of weeds in crops. Weeds in crops often must be controlled when still seedlings or just as they grow out of the seedling stage. If not controlled during this early growth, the crop may be lost due to the lack of effective control procedures for the mature weed in that particular crop.

At the beginning, approximately 3000 plant samples were identified during a year. Almost 20 years later, the requests totaled approximately 8000 identifications per year, one for every 15 minutes of a working year. Samples coming into the office were mailed or actually carried by someone. The size of the pile of boxes, envelopes, bags, and plants varied from about a foot to several feet per day. Difficulties in obtaining adequate samples coupled with years of identification experience led to a skill at identifying bits and pieces of plants. Sometimes the bits and pieces of the samples were all that were available. The fragments could be from physicians or veterinarians who were faced with a person or an animal having eaten it, or a piece of the plant left in a wound. Agricultural weed samples were frequently seedlings. Homeowners would generally grab a single leaf. Tightly wrapped plant samples sent via mail in the Gulf Coastal Plain states of the USA can arrive after a few days of humid, hot conditions as green goo. In an effort to be of help, each sample was carefully examined for any characteristic that could lead to identification. This effort was rewarding both for the information the senior author gained and for the help it provided those sending the sample. Forensics was not a part of the job description.

Law enforcement in the 1960s and early 1970s had very few cases involving plants as evidence, or, perhaps, few cases where they recognized that plant evidence was present.

When faced with such evidence the normal course of events was to contact the local county agent or a nearby forester. The United States has a system, initially to aid agriculture, that features an agricultural specialist in each county. Additionally, companies engaged in forestry are spread throughout the USA and the world. During that time county agents and foresters had limited to good skills at identification for the plants with which they usually dealt. Realizing their limitations, these few initial cases were referred to the Plant Identification and Information Service. At first the law enforcement cases totaled about three or four per year and most were concerned with civil matters. As the load increased so did the number of criminal cases. Oddly, for a period of time in the early years, many cases were concerned with attempting to find the crime scene for sexual assaults where the victim had survived, walked away, found help, and could not remember where they had been attacked. Case loads, both civil and criminal, have recently increased.

In 1986 Dr William Maples, a noted forensic anthropologist, who had requested help on several occasions, asked if the senior author was interested in joining the American Academy of Forensic Sciences. On being admitted to the Academy in 1987 and attending the first meeting, an amazingly wide range of disciplines and expertise was found to be available. At first the senior author was the only botanist in the Academy. Realizing that forensic botany was little utilized, he started giving lectures. In 1988 Dr Maples and the senior author tried, without luck, to get a course segment dealing with forensic anthropology, botany, dentistry, and entomology placed into the curriculum of the University of Florida College of Law. In 1991 forensic botany was permitted in a trial class lecture series in the College of Law. A symposium, Forensic Botany: Plants and Perpetrators, was organized for the American Institute of Biological Sciences Annual Meeting in 1988. While popular with local law enforcement and the press, his fellow biological scientists were much less than enthusiastic.

Also in the 1980s, Dr Michael Olexa, a Ph.D. plant pathologist and lawyer, was on a University of Florida sponsored circuit lecturing to help biologists give testimony in legal cases. Dr Olexa asked for help providing forensic botany presentations to the same groups. Lectures at the 1988 annual meetings of the Entomological Society of America and the Rocky Mountain Conference of Entomologists provided introductions to several entomologists who were doing research to help determine time since death using insect succession. Dr Lamar Meek of Louisiana State University was actively engaged in forensic entomology and provided introductions to other biologists who had an interest in forensics. These contacts encouraged further involvement.

In 1993 and in succeeding years, Dr Wayne Lord, an entomologist and FBI agent, provided an opportunity to lecture on botanical evidence and provide field exercises during a one-week training session for FBI agents and other law enforcement personnel. The course, Detection and Recovery of Human Remains, was conducted at the FBI National Academy at Quantico, Virginia. The enthusiastic level of interest from the agents was encouraging. This unique complex is connected by enclosed walkways and tunnels, leaving one to understand how some rodents feel. Other activities are always underway at the Academy so that, after the first startled response, kidnappings, arrests, bank robberies, car chases, gun shots, etc., seemed almost normal. Field sites were set up to show common types of plant evidence. Lectures emphasized the uses of this evidence. The senior author learned a great deal about surface remains and burial sites from his fellow lecturers. This particular course was discontinued several years ago, but those that attended still remember the lectures and continue to ask questions. This FBI course has inspired other institutions to offer similar training.

Training through courses for interested parties is crucial for the advancement of forensic botany. A three- to five-day Bugs, Bones, and Botany short course has been offered since 1999. Lectures and talks, while informative and entertaining, do not provide the level of information needed to process a scene. However, these lecture presentations show how plant evidence can be helpful with the aid of a professional. Botanists scattered around the world have been involved in a case or two, but only a few have worked on several to many cases. Except for biologists who work with DNA, most botanists, with the exception of two others, who have worked on forensic cases, have not been interested in joining the Academy of Forensic Sciences or any other forensic organization. To date no standards for training or expertise have been established for professional certification.

1 Introduction to forensic botany

David W. Hall, Ph.D.

Forensic botany is the study of plants and how they can relate to law and legal matters. Botany, while widely known as a science, has few professionally trained botanists. In proportion to the number of students trained in most other scientific disciplines, botanists are but a tiny fraction of the total number of individuals working in the field of botany. Many people who teach botany at two- and four-year colleges have perhaps taken only a course or two, and that is likely only because botanical training is typically included in a basic science curricula at the undergraduate level. Some college courses are combined zoology/botany courses, and as a result many college graduates have only portions of a full botanical course and never an entire course. Often, members of various professional plant societies (native plants, garden clubs, and nature organizations), environmental agency employees, and industry workers do not have any formal botanical education. With this low level of academic exposure it is no wonder that so few individuals understand the importance of plants, especially in criminal investigations.

Law enforcement officers and attorneys are no more informed about the science of botany, on average, than the general population. Therefore, important plant evidence is frequently overlooked. Sometimes this evidence can place a person or object at a crime scene, verify or refute an alibi, help determine time since death, the time of a crime, the place where a crime occurred, cause of death, or reason for an illness.

Botanical evidence in legal investigations

To be of value, plant evidence must first be interpreted by a botanist. Typically, legal investigators should seek a botanist with well-rounded training and experience who possesses knowledge of the various specialties within the botanical field in question (Figure 1.1). Some of the various botanical specialties involve systematics (plant names), anatomy (plant cells), morphology (plant structures), ecology (relationship of organisms within the environment), and physiology, chemistry, and genetics (DNA).

Forensic Botany: A Practical Guide, First Edition. David W. Hall and Jason H. Byrd.
© 2012 John Wiley & Sons, Ltd. Published 2012 by John Wiley & Sons, Ltd.

Figure 1.1   Dr David Hall, forensic botanist discusses the collection of plant evidence with students at a North Carolina State University forensic science workshop. (Photograph courtesy of Gary Knight.)

Not all botanists have such training. As in many fields, most professional botanists have specialized in one particular area, and have only scant knowledge concerning the forensic possibilities within others.

Plant systematics is the field that deals with the biological classification of plants. A plant systematist deals with the taxonomic classification of plants and utilizes standard nomenclature techniques to derive the scientific names of plants. Consulting with an expert in plant systematics is usually an excellent place for the legal investigator to start when trying to obtain expert assistance. Some experts work with only a single group of plants, but many, if not most, systematists work with many groups and have a good general knowledge of the entire field of botany. In order to place a name on a previously unknown plant, the existing information about that plant and its relatives must be reviewed. To have the capability to review this wide-ranging data, the plant systematist has taken courses and often conducted research and training in all of the areas of botany. They are by no means an expert in each of these fields, but the systematist usually does have connections with professionals who do have such expertise.

Legal plant definition

According to Webster's *Unabridged Dictionary*, ordinarily a plant is any living thing that cannot move voluntarily, has no sense organs, and generally makes its own food by photosynthesis. It is a vegetable organism, as distinguished from an animal organism, and there are many kinds of plants, exhibiting an extremely wide range of variation (Figure 1.2).

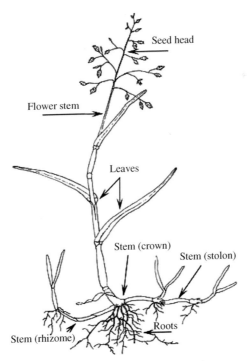

Figure 1.2 Diagram of the major parts of a common plant. (Courtesy of the Center for Turfgrass Science, College of Agricultural Sciences, The Pennsylvania State University.) The Cool-Season Turfgrasses: Basic Structures, Growth and Development.

In the United States, the legal definition of a plant can have sentencing ramifications. For instance, by legal definition a plant must have a connected stem and root. If the stem and root are separated, each becomes a plant fragment. Of course, not all plants have roots, but most do. It is illegal to possess plants or plant fragments of marijuana (*Cannabis sativa* L.), a plant with psychological and euphoric qualities (Figure 1.3). The US government measures the marijuana by either weight or the number of plants. Marijuana plant samples without roots are considered to be plant fragments and are weighed with other pieces, and a plant with a root can be counted as an entire plant regardless of size. Federal guidelines recommend sentences based on either the number of entire plants or various amounts by weight. If entire plants are found along with loose fragments that have been harvested for use or sale, the number of plants can result in longer sentences than the loose fragments, even if the plants are tiny and the weight is insignificant.

Botanical evidence in legal investigations

Not all legal matters are concerned with crimes against persons. Local and national ordinances can prohibit growing, selling, or importation of certain plants, such as

Figure 1.3 Marijuana leaf (*Cannabis sativa* L.), a plant which the United States government considers to be a controlled substance. Penalties for possession can be imposed based on total weight or number of plants. (Photograph courtesy of the United States Fish and Wildlife Service.)

marijuana. Identification of these plants is crucial for law enforcement. Examples of plants needing biological identification are plants listed as illegal drugs, dangerous agricultural weeds, and exotic species known to invade and destroy native habitats. A very large responsibility at all points of entrance to countries is that of illegal plant identification. All imported goods should be inspected for the many types of illegal plants and their parts. Suspected importation of harmful drugs, prohibited medicines, invasive weeds, including the seeds of prohibited plants (Figure 1.4), all need to be verified before interdiction can occur. Illegal plants are often found as fragments or as seeds or seedlings in shipments of other products such as grain or ornamental plants. Almost no-one can absolutely identify all the various plants that travel around the world with humans, and the identification of the various parts of those plants is even more difficult. Often such identifications go beyond the scope of law enforcement and border protection, and specialists are often needed.

While technically not plants, fungi have historically been studied by botanists. In many areas, naturally occurring plants, such as certain mushrooms (species of *Psilocybe*, the magic mushrooms), Peyote cactus (species of *Lophophora*), even marijuana when grown for fiber, are not illegal unless collected and prepared for use of their hallucinogenic properties. In fact, the cultivated form of marijuana used for fiber has very little hallucinogenic effect. However, unknowing individuals attempting to collect these plants often trespass on private property. Additionally, the mere transportation of plants used for

Figure 1.4 Marijuana seeds (*Cannabis sativa* L.) are often not recognized by law enforcement personnel due to the the their similar appearance to many other plant seeds. (Courtesy of Dr D. W. Hall.)

their hallucinogenic effects can be illegal. While rare plants are not protected from destruction on private property, transportation of them without proper permits is a violation of international, federal, state, and local laws. The rational for the regulation of the transportation but lack of penalties for the destruction of any protected plants naturally occurring on your property stems from old English law. In essence, if it grows on your property, you own it, but transportation on public roads can be regulated.

Alibis

In many criminal events, an alibi can prove to be very useful. Fortunately, very few individuals who are not botanists can successfully lie about plant evidence. Although a particular plant can grow in a wide range of areas, linking a plant to a particular spot or area can support or refute an alibi by showing that the person is, or is not, telling the truth. Because plants grow in specific areas, an object or person can often be linked to that location by plant evidence. Plant pieces found underneath a vehicle or on clothing can be linked to the location of the crime (Figure 1.5). Plant evidence can also show where the person or vehicle had previously been. Several locations can be combined to show movement relevant to the crime, and as a result victims and suspects can easily be connected to a crime scene.

Timing

Plants can give an indication of when an event occurred. Many plants are annuals, such as warm season crops like corn and soybeans. Some annuals are quite short-lived,

Figure 1.5 Vegetation found on suspension under suspect vehicle. (Courtesy of Dr J. H. Byrd.)

completing their life cycles in a matter of days or weeks, especially those in areas where the growing season is very short. These short-lived annuals may germinate as soon as the ground is warm enough and die when the weather becomes hot. Annual plants at high altitudes commonly are quite short-lived due to the very short periods of warm temperatures. Thus if a short-lived plant is present, the person or object would have been where that plant grew during its growing season.

Other characteristics also indicate seasonality and therefore can be of help. Plants that lose their leaves for the winter or in dry seasons will then regain leaves in warmer or wetter times. If a branch with no leaves attached is potential evidence, the branch could represent a time period during the winter or dry season. Leaves, flowers, and fruits that have fallen, sometimes in masses, can be of great use if the time when they fell or the amount of time it takes them to disintegrate can be determined. Thus, a single flower, fruit, or leaf may indicate something entirely different than a mass of them.

In areas of regular leaf fall the layers of leaves can indicate periods of time (Figure 1.6). Colder climates and very dry climates will normally accumulate several layers of leaves because the leaves take several years to decompose. In warm to hot climates leaves decompose very rapidly and may represent time intervals of only a few weeks or months. Leaves falling into standing water or buried in certain soil types are very slow to deteriorate due to the lack of oxygen, although they may be colonized by organisms that can speed up the process.

Figure 1.6 Layers of leaf fall can help to indicate the time period that has elapsed since the deposition of evidence (Courtesy of Dr D. W. Hall.)

Time of death (time since death, post-mortem interval) can be estimated by the use of plants. Tree rings are one type of plant evidence being used for time of death estimation (Figure 1.7). If a woody plant with seasonal growth producing annual tree rings is found growing on a grave or growing through a skeleton, the annual rings can be counted. The number of rings, corresponding to a year per ring, can show that the grave or skeleton was at this location for at least that many years.

Figure 1.7 Tree rings are one type of plant evidence used for time of death estimation. (Courtesy of Dr J. H. Byrd.)

Figure 1.8 The amount of wilt present on broken or damaged leaves and branches can help to establish a time frame. (Courtesy of Dr James L. Castner.) (Please refer to the colour plate section.)

The time of an event can be determined using other plant characteristics. A broken branch usually will wilt. The amount of wilt can be determined experimentally if the evidence is photographed and collected correctly (Figure 1.8). The time for sap to dry after a branch is broken or bark is nicked can also be determined by experimentation. Leaf wilt can also occur when plants are uprooted or the root systems are severely disturbed. Determining the approximate time the sap has been accumulating or the leaf wilting depends on having a botanist at the scene quickly. As with most types of physical evidence, proper photography is essential. The botanist will need to determine the identification of the vegetation and conduct an experiment by breaking similar branches. By examination at regular intervals a botanist will be able to determine the relative time it takes the sap and/or wilt to match that shown in the photographs. For best results, several replications are necessary to get an average time period. The investigator should keep in mind that if conditions change between the time of discovery and when the experiment is started, the results may not be applicable. If parts of plants that are green (have photosynthesis) become buried, they will gradually lose their green color as the chlorophyll breaks down from the lack of sunlight. The color will gradually become yellow and eventually brown (Figure 1.9).

The difference between the shades of yellow and the natural shade of green can be determined by burying many replications of the same plant parts at the scene and uncovering them at predetermined intervals. When the experimentally buried plant color matches that collected at the crime scene, it can roughly determine when the plant part was buried. This experimental process can be utilized to determine when the body or object was put in the ground. This time period may, or may not, indicate the time of death. It should be kept in mind that the suspect could have killed the victim some time before the body was buried. Also, at times bodies are moved and this experiment

Figure 1.9 Chlorophyll degradation of plant leaf from a burial site can be used for time interval estimations. (Courtesy of Dr J. H. Byrd.) (Please refer to the colour plate section.)

procedure can be used to show the approximate time of the most recent burial. Likewise, it could show, after the body was moved, when the original burial took place. To be effective this experiment must be done under the same conditions at the scene where the plant material was discovered. The sooner the experiment can be started the more accurate the findings will be. Photographs of the yellowed plant parts must be of very high quality. The botanist must quickly match the vegetation and bury it, and each replication must be photographed at the time of burial and at the time of retrieval. As in any experiment, the more replications per time period, the better the results.

Within the United States several states have legally binding wetland rules that prohibit disturbance of wetlands. Generally, these rules are a civil matter. Wetland boundaries are determined using plants, soils, and hydrology. Disputes frequently arise concerning the boundary locations and prohibited activities in wetlands. The original location of a wetland and the time during which the suspected violation occurred can be an issue solved by plant evidence. Tree rings can provide the year of the original disturbance and other buried plant parts may provide the season, such as finding buried flowers and knowing at what time of year the blooms occurred.

Gravesite growth

A common fallacy is that the vigorous growth of plants on and around a gravesite or body is due to the nutrition provided by the deterioration of the body. In the early post-mortem period this is simply not true. First, a body placed on plants will keep sunlight from reaching the plant, causing most plants to die due to a lack of photosynthesis. An exception would be plants connected by extended stems. The above-ground portion of the plant under the body would die due to the shading, but the extended stem can be

Figure 1.10 Area of decomposition and the resulting lack of vegetation. Over time, plant growth will reoccur. (Courtesy of K. Shaw.)

fed by the part away from the body. Second, fluids left by deterioration are caustic to plants. Most stems and root systems would eventually die because of the caustic fluids and lack of photosynthesis. Also, the fluids will prevent other plants from colonizing the soil (Figure 1.10).

Figure 1.11 Vigorous plant growth over a grave site due to soil disturbance from a burial. (Courtesy of Dr J. H. Byrd.)

The easily observed vigorous plant regrowth over and around a gravesite is due to the disturbance of the soil, not any nutrition provided by the remains (Figure 1.11). The soil disturbance exposes seeds in the soil to environmental factors that will trigger growth. Germination of seeds is often started by changes in temperature, water, or light, all factors that can be kept away by compacted soils or depth of the seeds. Almost all soils contain a seed bank. A seed bank consists of all the seeds that have fallen on or been transported to the area and remain buried. Soil disturbance frequently brings these seeds to the surface. Weedy species are quite common. A simple experiment, often set up in entry level ecology courses, places a square of soil a few inches thick in a pan with a bit of water to await the germination of the seeds. Varying results can be found by subjecting the soil to prescribed differences in temperature, water, and light. If the square of soil is removed in an undisturbed condition the results are quite different from soil that is removed and mixed.

Stomach contents

Plants within stomachs or feces can be identified by means of their anatomy. Bock *et al.*[1] (1988) detail identification of plant food cells in gastric contents. Surprising deductions can occasionally be made on the basis of ingested food. In one case an analysis of the stomach contents of two murdered women led to the assumption that a serial killer was responsible. Sometimes the contents can lead investigators to the place of a last meal. Ingestion and partial digestion of a poisonous plant could also be determined. In some regions of the world plant poisoning is more frequent. Most poisoning must be determined by chemical analysis.

Summary

Very few professionals involved with law enforcement have a background or training in botany. Botanical evidence can be very important as it may be possible to check the validity of an alibi, place someone or something at a particular location, help determine the time since death, or assist with timing of other events. There are many kinds of botanists who can help with the interpretation of plant evidence, such as plant chemists, DNA experts, or those that deal with stomach or feces contents. Plants associated with a crime as well as illegal plants such as those with serious psychological or euphoric effects, or plants that are prohibited because they are dangerous agricultural weeds, or exotic weeds that invade and destroy our native habitats all need classification and diagnosis by a plant identification specialist for proper disposition. Botanists are also essential for certain legal dilemmas regarding environmental rules and regulations.

---

[1] Boch, J.H., Lane, M.A., and Norris, D.O. (1988) *Identifying plant food cells in gastric contents for use in forensic investigations: a laboratory manual.* US Department of Justice, Washington, DC.

2 Plants as evidence

David W. Hall, Ph.D.

Years of teaching botany has shown that many, if not most, crime scene personnel have a very difficult time finding plant evidence. Often, investigators do not see a plant because they do not recognize the object as a plant. Recognition of the various common groups of plants is necessary to be able to find and utilize plants in forensic investigations. There is no need for the crime scene investigator to identify the types of plants, just to recognize that the particular evidence is a plant or is from a plant in order to collect it properly.

Types of plants

Vascular plants

Vascular plants are those that have a transport system to conduct water and food throughout the plant, analogous to the vascular system in animals. Roots function to anchor the plant to the substrate and provide mineral and water uptake. The shoots may be branched or unbranched, and function in the conduction of food, water, and minerals throughout the plant. Leaves absorb light and through the process of photosynthesis produce food for the plant. (Figure 2.1)

Seedless plants

Ferns   Most people are familiar with the leaves of ferns as fronds, and these can be divided or undivided. The stems are frequently underground, but can grow upright. The backs of the fronds contain reproductive material known as spores, which may appear as small dots in groups, lines, or masses. The spores are very tiny and appear in mass as a powder (Figures 2.2–2.4).

Forensic Botany: A Practical Guide, First Edition. David W. Hall and Jason H. Byrd.
© 2012 John Wiley & Sons, Ltd. Published 2012 by John Wiley & Sons, Ltd.

Figure 2.1 Vascular system of a bramble leaf (Family: Rosaceae; Genus: *Rubus*). (Courtesy of Wikimedia Commons and Richard Wheeler. www.richardwheeler.net.)

Figure 2.2 Vascular seedless plants, such as this Boston fern, are often used in landscaping as ornamental displays. (Courtesy of Dr. J. H. Byrd)

Figure 2.3 The leaves of ferns are termed fronds and the stems often grow underground, as with this spreading tri-vein fern. (Courtesy of Dr. J. H. Byrd)

Figure 2.4 Some common ferns are very small and often go unnoticed. Southern grape fern is shown. (Courtesy of Dr. J. H. Byrd)

Figure 2.5   Cycads, such as this Emperor sago are often confused with palms. (Courtesy of Dr. J. H. Byrd)

Seed Plants

Conifers   This is a group of seed-bearing shrubs and trees. The seeds of this group are not enclosed, but are borne on the surface of scales of cones, as in pines, firs, cedars, and yews. Most conifers are evergreen and they are found throughout the world.

Cycads   These are tropical plants that resemble palms. The woody trunks bear stiff, fernlike, highly divided leaves in a cluster at the top (Figure 2.5).

Flowering plants

These compose the largest group of plants and can be both non-woody (herbaceous) and woody. The flowering plants are morphologically, rather than systematically, separated into monocotyledonous (monocots) and dicotyledonous (dicots) plants, and their seeds are borne inside a covering. Common examples are acorns, apples, grains, sunflowers, and citrus.

Monocots

These are grass and grass-like plants in which the embryos (seedlings) have one seed leaf when they emerge from the soil. They also exhibit parallel-veined leaves, inconspicuous flowers, flower parts in multiples of three, and no secondary growth. Common examples are grasses, sedges, rushes, palms, lilies, and orchids (Figures 2.6–2.9).

Figure 2.6   Grasses are common examples of monocot plants. The St Augustine lawn grass is shown. (Courtesy of Dr. J. H. Byrd)

Figure 2.7   Sedges, often confused with grasses, are monocot plants. (Courtesy of Dr. J. H. Byrd)

Figure 2.8 Monocot plants can also be large trees, such as this palm. (Courtesy of Dr. J. H. Byrd)

Figure 2.9 (a) Many palms do not produce tall trunks and are common ornamental plants. A monocot needle palm is shown. (b) Detail of a needle palm showing the needles that give this plant its name. (Courtesy of Dr. J. H. Byrd)

Figure 2.9   *(Continued)*

Dicots

These are broadleaf plants in which embryos (seedlings) have two seed leaves when they emerge from the soil. They also exhibit netted leaf veins, showy flowers, flower parts in fours and fives, and often have secondary growth. Common examples are hickories, oaks, beans, mints, and asters (Figures 2.10–2.14)

Figure 2.10   Dicots, like this live oak, are often large broad-leaf trees. (Courtesy of Dr. J. H. Byrd)

Figure 2.11   A dicot seedling showing the two seed leaves. (Courtesy of Dr. J. H. Byrd)

Nonvascular plants

Nonvascular plants do not have an organized transport system to conduct water and food. The most common nonvascular plant groups with which crime scene personnel will come in contact are mosses and liverworts.

Figure 2.12   Many dicot plants are widely cultivated and are popular because of their large and colorful flowers. (Courtesy of Dr. J. H. Byrd)

Figure 2.13   Lantana, a common weedy shrub, is a dicot and known for its hardiness and drought tolerance. It is common in urban and rural areas. (Courtesy of Dr. J. H. Byrd)

Mosses

A moss is a small green organism lacking true roots, stems, and leaves, although the plants appear to have stems and leaves. They may be erect or creeping and reproduce by capsules filled with spores (Figure 2.15).

Algae

Algae (singular: alga) are nonvascular plants which typically possess chlorophyll and are photosynthetic, but have no true root, stem, or leaf. The organisms may be composed

Figure 2.14   This dicot, a plumbago, is an ornamental shrub that is growing in popularity and commonly planted in urban settings. (Courtesy of Dr. J. H. Byrd)

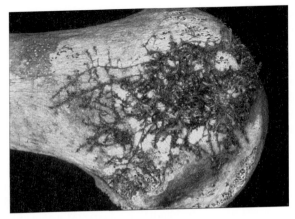

Figure 2.15   A moss growing on human bone can useful for determining a portion of the postmortem interval. (Courtesy of Dr. J. H. Byrd) (Please refer to the colour plate section.)

of one to many cells. Algae are very widely distributed throughout the world, especially green algae. They grow in wet to merely damp places, such as moist soils. Large masses of algae are often known as "blooms." Algae are put into several groups: blue-green, green, golden-brown, brown, and red. Blue-green algae are widespread in many environments, being especially common in marine and nitrate-rich habitats. Blue-greens are frequently noticed in polluted water as a slippery, smelly, blue-green to black, floating mass (Figure 2.16). Green algae are most often noticed as freshwater "pond scum." Golden-brown algae occur in the upper levels of both fresh and marine

Figure 2.16   The presence of algae can link individuals and items to particular bodies of water. A blue-green alga, commonly known as "pond scum", is shown. (Courtesy of Dr. J. H. Byrd) (Please refer to the colour plate section.)

waters. Diatoms are common representatives in both fresh and marine water, and sometimes form a "pond scum." Brown algae mostly occur in marine water. They are attached to a substrate and frequently have gas-filled bladders to keep the upper parts of the plants floating to take advantage of the light. Red algae are also found primarily in marine habitats. Red algae are typically attached to a substrate. They frequently occur at greater depths than other algae, to about 200 meters. Some common examples are seaweeds and pond scums.

Algae can present distinctive opportunities for linking a suspect to a location or determining a time of the year. An alga can be unique in a particular habitat or an assemblage of algae can be characteristic. Often particular combinations of algae can define a body of water. Some algae have very resistant coverings (diatoms) and are extremely resistant to decay, lasting thousands of years. Such long-lasting evidence can be of importance, but will need to be handled by an expert just as other trace evidence should be. Chapter 5 provides a complete examination of the importance of algae to the forensic sciences.

Nonplant groups traditionally studied by botanists

Fungi

Fungi (singular: fungus) form a group of organisms that lack chlorophyll, obtain nourishment from other sources, and reproduce by spores. These nonphotosynthetic organisms are most commonly thought of as mushrooms. Previously they were considered to be part of the plant kingdom but are now considered more closely related to animals. However, mushrooms are but one of the groups of fungi. The bodies of fungi are usually filamentous and are the agents responsible for much of the disintegration of organic matter. Many fungi have both under ground and above ground parts. The spores of fungi, like pollen, are microscopic and move by air currents, water, insects, and larger animals. Spores and the filaments of fungi can be of help to determine time of death and prior locations of the body. In addition to mushrooms, molds, mildews, toadstools, smut, and rust are typical types of fungi (Figures 2.17 and 2.18).

The rate of growth of fungi, as with many other plants at a scene, can be instrumental in determining the timing of events. For example, a collision between an automobile and a train occurred at a railroad crossing. During the investigation, it was discovered the stop sign at the crossing had been knocked down. Investigators were unsure if the sign was damaged as a result of the collision, or if the sign was down prior to the event. If the sign was up at the time of the collision, the railroad's liability, if any, would be much lower. A fungus was instrumental in determining how long the stop sign at the crossing had been down. The fungus had grown over the base of the broken sign post. The rate of growth for that species of fungus was known and it was used to demonstrate that the sign had been down long before the collision.

Figure 2.17   A fungus, commonly known as a "rust", on the trunk of a pine tree. (Courtesy of Dr. J. H. Byrd) (Please refer to the colour plate section.)

Lichens

A lichen is a cooperative arrangement between algae and fungi. Recent discoveries are showing that several to many fungi are involved in each individual lichen, and often more than one alga is included. Sexual reproduction is by the fungus. The colors of the organisms are very variable and often bright. Several growth forms occur: crustose,

Figure 2.18   Shelf fungus, common in forested environments, can be utilized to provide time intervals based on its rate of growth. (Courtesy of Dr. J. H. Byrd) (Please refer to the colour plate section.)

Figure 2.19 A crustose lichen is a community of algae and fungi. Lichens can be extremely valuable in creating time intervals based on their known growth rates. (Courtesy of Dr. J. H. Byrd)

foliose, fruticose, and gelatinous. Crustose lichens exhibit a thin crust tightly adherent to a substrate and cannot be removed without tearing it apart. Foliose lichens resemble small, brittle, crusty, wavy leaves, and can be removed from a substrate intact. Fruticose lichens are branched, erect, or hanging, and are attached at one end. Gelatinous lichens are black or bluish-black gelatin-like blobs (Figures 2.19–2.21).

Figure 2.20 A common foliose lichen, which is known to grow on any porous surface under the proper environmental conditions. (Courtesy of Dr. J. H. Byrd)

Figure 2.21   A fruticose lichen is branched and easily removed from items. (Courtesy of Dr. J. H. Byrd) (Please refer to the colour plate section.)

Plant habitats and associations

Ecology

Plants grow with other plants in associations. The communities in which plants grow are governed by soil, temperature, wind, moisture, altitude, latitude, and longitude. The specific environment in which an organism lives is called a habitat. Forests and deserts are two easily recognized plant habitats. Some habitats are widespread, such as the northern coniferous forest. Other habitats are narrowly limited, such as the scrub found in central peninsula of Florida in the United States (Figure 2.22). Obviously, finding plant evidence from an unusual or rare habitat can be extremely valuable to a legal investigation.

Figure 2.22   A typical Florida scrub habitat. Such habitats, restricted to certain geographic areas, contain many hundreds of species that can be used to link individuals and objects to particular locations. (Courtesy of Dr. D. W. Hall)

All habitats are dominated by certain species. The name of the habitat is usually determined by the dominant plants. This domination can be by numbers of individuals of that species or by the size of the species. In grass prairies the dominant species are the vast numbers of certain grasses. In a cypress swamp the dominant plants are the very large cypress trees, although there may be fewer cypress trees than the many other smaller plants present.

Usually the habitat name is an indication of the dominant types of plants and the substrate on or in which they are growing. For instance, swamps have standing water and trees, prairies are open grass lands, flat woods have trees and little elevation, tundras are flat, boggy, treeless plains at a high altitude, forests are covered with trees, savannas are grasslands with scattered trees or clumps of trees, deserts have few plants and little rain, and ditches are elongated man-made trenches. Other descriptive words can locate the habitat in a particular climate, such as tropical, temperate, or northern.

Most plant associations have been investigated and described by botanists. Identifying any particular plant known to be an element of the particular habitat can indicate other species in the association. Much additional information can be gained from a plant ecologist (one who studies plants and their relationship to the environment) based on relationships among plants. Determination of even a single plant can be of great help in placing a person or object at a location.

Lack of habitat

Many species of plants can occur in more than one habitat, for example various weeds can be common in any cultivated or otherwise disturbed area. This may mean that a specific habitat cannot be indicated. Still, some kinds of information can be useful even if a specific plant association is not indicated. For instance, one might be able to determine if the area is wet or dry, or at least if the plant material is from a different area than that in which a body was found (Figure 2.23). An example of a widespread weed growing in many habitats is creeping beggar-weed (*Desmodium incanum* DC.) (Figure 2.24). Creeping beggar-weed occurs in almost every open or disturbed dry habitat in its range. It does not occur in wet areas so it can prove that someone was in a dry habitat.

Plant characteristics/plant morphology

Plants are combinations of various structures such as roots, stems, branches, leaves, and flowers. Many plants have unique parts that can be used by a systematic botanist (one who studies plant names) to determine its name. Different kinds (species) of plants have leaves of various shapes and colors as well as unique tips, bases, and margins. Leaves can be arranged in distinct patterns and may or may not have stalks. Likewise, stems and their branching patterns can differ substantially, and root systems vary tremendously.

The principle means by which scientists separate plant species is by their flowers and fruits. However, these plant parts are not always found at the crime scene, and other

Figure 2.23 Typical dry, open, weedy, mowed field in which creeping beggar-weed (*Desmodium incanum* DC.) can be found. (Courtesy of Dr. J. H. Byrd)

plant characteristics must be used for identification. Without flowers or fruits, identification can be more difficult and less certain. Many characteristics other than flowers or fruits, such as hairs and scales on leaves, are microscopic and cannot be seen without magnification (Figure 2.25).

Figure 2.24 Creeping beggar-weed (*Desmodium incanum* DC.) easily attaches to clothing and animal fur. Its breakaway parts can used to link individuals and animals to certain habitats and specific geographic locations. (Courtesy of Dr. J. H. Byrd)

Figure 2.25 Trichomes, or plant hairs, serve a variety of functions from deterring feeding to altering the micro-environment around the plant leaf. (Courtesy of Dr. J. H. Byrd)

The usefulness of the name of the plant should be self-evident: the name of the plant is the index to all information concerning that plant! It is the key that unlocks the door to plant identification. Very little information can be found without the plant name, and practical experience, the library, and various experts are the best resources for information about plants.

Basic plant characteristics for the forensic investigator

To aid crime scene personnel, a very brief summary of plant characteristics is provided in this chapter.

Many years of training and creating mock crime scene scenarios for teaching plant evidence retrieval have shown that the overwhelming majority of plant evidence is often missed. Most personnel, if they recognize the material as being from a plant, do not recognize any particular plant as being different from another. The following characteristics should be learned to provide a background for recognizing the different kinds of plant evidence at a scene.

Habit

In botany, a habit can be described as the tendency of a plant to grow a certain way. One major habit is that plants can be primarily woody or herbaceous (not woody).

Woody plants

Woody plants are those with hard fibrous cells beneath the bark and they are usually classified as trees or shrubs. The difference between these two groups of plants is that trees normally have a single stem and shrubs have multiple stems (Figure 2.26). Many

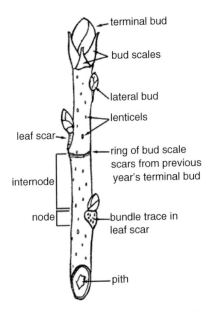

- terminal bud
- bud scales
- lateral bud
- lenticels
leaf scar
- ring of bud scale scars from previous year's terminal bud
internode
node
- bundle trace in leaf scar
- pith

Figure 2.26 Stem of a typical woody plant. (Courtesy of Clemson University Extension.)

plants with woody stems have non-woody branches. Woody plants are normally perennial and can live for a few to several hundred years (Figure 2.27).

Herbaceous plants

Herbaceous plants are those that lack a persistent woody stem. The large majority of plants are herbaceous. Herbaceous is not to be confused with the term herb, which refers to plants used commonly for medicine, seasoning, or food (Figures 2.28 and 2.29).

Figure 2.27 Methuselah tree, thought to be one of the oldest plants on earth. (Courtesy of Fotopedia, Creative Commons – Mike Crebis.)

Figure 2.28   A herbaceous plant in flower. (Courtesy of Dr. J. H. Byrd)

Stems

Stems are defined by having a bud in the axil of a branch. Stem shapes can be round, square, or have one or two sides flat and the other sides rounded, concave, or convex. Stems can be above or below ground. Above-ground stems can be erect, leaning, or run along the surface. Below-ground stems can be upright or run parallel to the surface. Some

Figure 2.29   A herbaceous stem tip. (Courtesy of Fotopedia – Creative Commons – John Schilling.)

Figure 2.30 Knowing the shape of plant stems can assist in identification of the plant and a description of the stem should always be included in the written notes made about the plant on collection. (Courtesy of Dr. J. H. Byrd)

below-ground stems are modified into storage structures such as a tuber, which is a short thick underground storage organ with "eyes" (as in a potato), which are the leaf scars at each node, like those that appear on almost any stem. Below-ground stems may also possess a corm, which is the swollen underground tip of a stem (Figures 2.30 and 2.31).

Leaves

A leaf is an above-ground plant organ specialized for the process of photosynthesis. As a result, leaves are typically flat (laminar) and thin. This shape maximizes the surface area exposed to light. Most frequently we notice the leaves on plants that are not in flower. Leaves of different kinds of plants, while variable within the species, have distinct

Figure 2.31 A rounded stem is the most common stem shape. However, the investigator should not assume that all stems are round. A detailed inspection should be made and supplemental photographs taken to aid in identification. (Courtesy of Dr. J. H. Byrd)

Figure 2.32   A vine showing an opposite leaf arrangement. (Courtesy of Dr. J. H. Byrd)

characteristics that can enable a botanist to definitively name the plant. Each leaf is composed of a blade and most often a stalk. Being aware of common leaf characteristics is very important for investigators involved with the collection of evidence. All crime scene personnel should be able to generally categorize leaves that appear different. The basic characteristics useful for separating different plants based on leaves are described and shown below.

Leaf arrangement

Leaf arrangement on stems can be opposite, alternate, or whorled. The great majority of plants have alternate leaves. Since fewer plants have opposite or whorled leaves, these arrangements may lead to a quick identification (Figures 2.32–2.34).

Leaf types

Simple   A simple leaf is not divided into parts and has a single blade. A simple leaf may be lobed. The divisions between the lobes of a simple leaf do not reach down to the midrib. This single blade may or may not have a stalk (Figure 2.35).

Divided   A divided leaf is a leaf that has more than one blade and may or may not have a leaf stalk. A leaf can be divided once, twice, or three times. If more than one blade is present each blade is called a leaflet. The leaflets may or may not have stalks (Figures 2.36 and 2.37).

Leaf margins

The margin of a leaf refers to its edge. While leaves may differ in size on any one plant, they will all possess the same general shape and margin. Therefore, the leaf margin can

Figure 2.33 A common ornamental plant exhibiting an alternate leaf arrangement. (Courtesy of Dr. J. H. Byrd)

be an indicator of the species of plant in question. The margin of each blade may be smooth or have teeth. There are many kinds of teeth: straight, bent forward, blunt, sharp, double, large, small, etc (Figures 2.38 and 2.39).

Figure 2.34 A whorled leaf arrangement, one of the least common types, can be used to quickly identify many plants. (Courtesy of Dr. J. H. Byrd)

Figure 2.35   A lobed simple leaf with a very short stalk attaching to a stem. (Courtesy of Dr. J. H. Byrd)

Figure 2.36   A once-divided leaf with three leaflets and a long stalk. The leaflets also have stalks. (Courtesy of Dr. J. H. Byrd)

Figure 2.37   A once-divided leaf with many leaflets. These leaflets do not have stalks. (Courtesy of Dr. J. H. Byrd)

Figure 2.38 Leaves with a serrate margin. (Courtesy of Dr. J. H. Byrd)

Leaf lobes

The number of lobes on a blade is important for identification. Lobes can be regular or irregular in shape and number. Lobes often have teeth on the margins. The lobes can be very difficult to separate from larger teeth (Figure 2.40).

Leaf tips

The apex of a blade can be obtuse (rounded), acute (slightly pointed), acuminate (pointed), bristle tipped (apex sharply pointed), truncated (flat), or heart-shaped (notched) (Figure 2.41).

Leaf shapes

Common blade shapes are elliptical (like an American football), round, lance-shaped (broadest at the base), oblanceolate (the opposite of lance-shaped, broadest at the tip), oblong (sides are parallel), heart-shaped, oval (like an egg), crescent (like a waning or waxing moon), awl-shaped (tiny with a sharp point), and many others that are less frequently encountered (Figure 2.42).

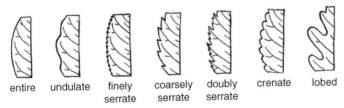

entire   undulate   finely serrate   coarsely serrate   doubly serrate   crenate   lobed

Figure 2.39 Examples of several types of leaf margins. (Courtesy of Clemson University Extension.)

Figure 2.40   Lobes on leaves may be elongated. This is leaf shape is considered to be five-lobed and has teeth on the margin. (Courtesy of Dr. J. H. Byrd)

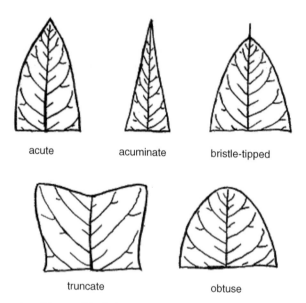

Figure 2.41   Examples of types of leaf tips. (Courtesy of Clemson University Extension.)

Hairs

Plant hairs (trichomes) frequently occur on various parts of the blades and stalks. Many kinds of hairs occur. More than one kind of hair can occur on the same part of a plant. Hairs may be simple with one cell, many celled, long and wispy, stiff and erect, branched, or colored. Hairs can be located on any part of the leaves or stems. Hairs can

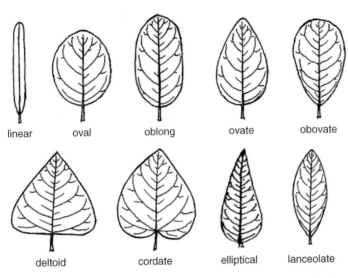

linear   oval   oblong   ovate   obovate

deltoid   cordate   elliptical   lanceolate

Figure 2.42   Examples of leaf shapes. (Courtesy of Clemson University Extension.)

appear to be in straight lines or cover an entire surface. Hairs often occur along veins or margins (Figure 2.43) and have many functions. In some instances, plant hairs be stiffened and thus interfere with feeding on the plant by small and large herbivores. Dense hairs on plants help to insulate it and keep frost away from the living surface cells. In arid environments plant hairs slow the flow of air across the leaf surface, thereby reducing evaporation. Dense coatings of hairs may also reflect solar radiation or collect moisture from morning dew and rainfall.

Figure 2.43   Trichomes, or plant hairs, densely covering the upper surface of a leaf blade. (Courtesy of Dr. J. H. Byrd)

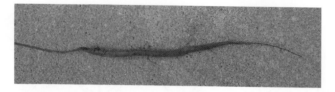

Figure 2.44   The large central root of a tap root system. (Courtesy of Dr. J. H. Byrd)

Roots

Roots are often an important aid in the identification of a plant, but since roots are usually located below the soil surface they go unnoticed and underutilized by legal investigators. Morphologically, roots differ from stems in that they have no bud in the axil of a branch. The two most common types of roots are fibrous and tap. Fibrous root systems spread outward from the base of the plant with larger branches dividing and turning into many smaller branches, and ultimately into very fine roots. A tap root system has a large central root, usually growing downward, with a few smaller branch roots. While individual roots may be very difficult to identify, the type of root system can be of great help (Figures 2.44 and 2.45).

Flowers

Flowers are one of the primary means by which plants are identified. Flowers are commonly composed of sepals, petals, stamens, and pistils. Sepals are collectively called the calyx. Starting on the outside of a flower, sepals are the first layer of the

Figure 2.45   Fibrous roots spreading from the base of a plant. (Courtesy of Dr. J. H. Byrd)

Figure 2.46 Anthers covered with pollen. (Please refer to the colour plate section.) (Courtesy of Dr. J. H. Byrd)

structure of a flower. Usually sepals are green and occur in the same number as the petals. Petals are collectively called the corolla, and they are the second layer of structure in a flower. Petals are often brightly colored. Stamens, the male part of a flower, are the third layer of structure and are located just inside the petals. The stamens have an anther and sometimes a stalk, and the anther is filled with pollen, which upon maturity will pollinate other plants of the same species (Figure 2.46). The pistil is the female part of the flower and is located at the center. It is composed of a stigma (the sticky tip that collects pollen) at the top, the ovaries at the bottom, and a style connecting the stigma to the ovaries (Figures 2.47 and 2.48).

Figure 2.47 Flower showing five petals surrounding many stamens and stigmas. (Courtesy of Dr. J. H. Byrd)

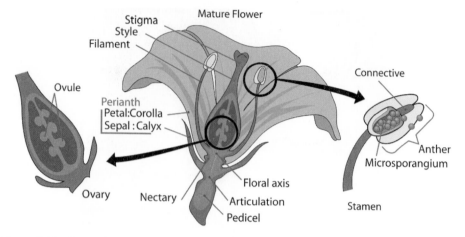

Figure 2.48 Parts of a typical flower. (Courtesy of Wikimedia – Creative Commons – Mariana Ruiz.)

Fruits

A fruit is a ripened ovary that provides the plant with the means for reproduction. Some fruits are hard and contain a single seed, such as oak acorns, hickory nuts, and coconuts. Some are fleshy with a single seed (plum and holly), while others are fleshy with several seeds (tomato, banana, and blueberry). Other seeds exist as a hard nutlet (sunflower and mint), a capsule with a few to many seeds (morning-glory, eucalyptus, and violet), an aggregate fruit composed of many small fruits (blackberry and raspberry), a dry to fleshy pod (bean and peanut), or are dry and winged (maple and elm) (Figures 2.49 and 2.50).

Figure 2.49 Fleshy fruits, each containing a single seed. (Courtesy of Dr. J. H. Byrd)

Figure 2.50 A hard fruit (acorn) contains a single seed. (Creative Commons- Fotopedia - Bryant Olsen)

Plant dispersal

Plants have both simple and complex ways of moving seeds and vegetative reproductive pieces. For reproduction to be most effective there must be a means to move the seeds some distance from the parent. Some fruits and seeds are edible so that animals ingest and carry them to new locations (Figures 2.51 and 2.52). Since most animals, including humans, eat fruits for nourishment this is a very effective mechanism for transport as seeds are often consumed with the fruit and can be deposited some distance from the source.

Figure 2.51 Some edible fruits, such as oranges, have multiple seeds. However, cultivated seedless varieties exist. (Courtesy of Dr. J. H. Byrd)

Figure 2.52 Some edible fruits, such as this mango, have single seeds. (Courtesy of Dr. J. H. Byrd)

Many seeds are light and have hairs and/or papery wings (Figure 2.53). Usually these are very small so that wind provides transport. Seeds and fruits with air spaces are adapted to be moved by water. Water can move fruits and seeds by providing a medium where they can float or be washed by currents. Smaller seeds and fruits can be moved short distances by water drops splashing.

Figure 2.53 Some plants, such as this red maple, have seeds with broad flattened areas which serve to transport the seed on the wind. (Courtesy of Dr. J. H. Byrd)

Figure 2.54 Some fruits are structured with spines designed to penetrate the hide or skin of passing animals or humans. (Courtesy of Dr. J. H. Byrd)

Flowers, fruits, seeds, and other pieces of plants can have hooks, hairs, spines, sticky secretions, and breakaway parts so they can easily attach or stick to passing animals, including humans. Animals commonly provide distribution by other means than attachment to their fur. Small fruits and seeds can be stuck in the soil on feet. In the case of humans the plant reproductive parts usually stick to clothing, but sandspurs with spines can easily penetrate the skin (Figures 2.54 and 2.55). Various

Figure 2.55 Other fruits have hairy bristles at the seed tip which are designed to become easily entangled in the fur or hair of passing animals. These seeds commonly adhere to clothing. (Courtesy of Dr. J. H. Byrd)

modern vehicles now provide more reliable, if often unintended, long-distance dispersal.

Many seeds and other plant propagules are very small, often being no more than a tenth of a millimeter in length or diameter. However, pollen can be much smaller (see Chapter 6). Visualization and identification of such small objects requires careful examination, often with magnification. A hand magnifying lens or a loupe is an excellent tool for the scene investigator.

On collecting any plant evidence, the two most important actions are to have a qualified botanist name the plant and have a qualified botanist interpret the evidence. Landscapers and gardeners, while having considerable plant information, are usually not qualified to analyse forensic evidence or give expert testimony.

3 Evidence collection and analysis

David W. Hall, Ph.D. and Jason H. Byrd, Ph.D.

Botanical evidence is most often preserved by simply pressing plant material and allowing it to dry naturally. This is because plants retain nearly all of their morphological characteristics after drying. Although the colors and shapes of the fleshy parts of plants, such as fruits, often change, these characteristics can be noted and recorded when they are collected. Photography is an excellent method to use for the documentation of the physical characteristics observed at the crime scene before the collection drying process begins. It is important for the investigator to understand that plant material does not have to be "green" to be useful. The best method to preserve evidence is in paper, pressed between sheets of newsprint, telephone book, catalog, or in an inexpensive commercially made plant press (Figure 3.1). Paper bags can be used for drying and storage, but there is a risk of shattering dried plant parts if the bag is handled roughly or crushed. Pasteboard boxes are excellent for drying larger pieces and their rigid structure helps prevent crushing.

When documenting and collecting fungi, it should be photographed in a manner that shows the color and shape of the fruiting parts. Fungi should be collected in pasteboard boxes, but if one is not available a paper bag will suffice until a box can be found. The special collection methods for algae and for plant DNA are described in Chapters 5 and 9.

Since botanical evidence is collected in paper, certain collection information needed for the chain of custody and location notes can be written directly on the paper or cardboard container (Figure 3.2). Documentation of the case number and item number is essential. However, additional information such as the date and time of collection (to the minute), and specific location information (exactly where found) must also be recorded for botanical evidence. Most other information can and should be recorded in the investigator's written notes and transcribed into a case file. The crime scene and

Forensic Botany: A Practical Guide, First Edition. David W. Hall and Jason H. Byrd.
© 2012 John Wiley & Sons, Ltd. Published 2012 by John Wiley & Sons, Ltd.

Figure 3.1 Commercially available botanical press. (Courtesy of Dr. J. H. Byrd)

evidence location information should include some reference to a major road, lake, or other documented landmark easily identified in the environment or located on a map.

Fresh plant material contains moisture and sugars. Both of these elements promote the growth of bacteria and fungi, and the higher the sugar and/or moisture content, the faster the decomposition and degradation of the sample. If heat is also present, complete decomposition of the plant sample can occur within 2–3 days, with many of the distinguishing characteristics disappearing within a few hours. To help prevent plant decomposition and sample degradation, it is crucial to not collect plant material in

Figure 3.2 Identifying information can be written on the paper used to dry the specimen. (Courtesy of Dr. J. H. Byrd)

Figure 3.3 Inexpensive portable refrigerators are extremely useful in the preservation of botanical evidence until the specimens can be properly pressed and dried. Many mobile crime scene units have these as standard equipment. (Courtesy of Dr. J. H. Byrd)

plastic bags or any non-porous container. However, if plant material is stuck to a body part or soaked in fluids it may be collected in plastic and refrigerated until it can be safely dried. Coolers with ice or chemical ice (cooling packs) can be used for temporary storage in the field. However, the plant material should not be placed directly on the ice to avoid freezing and damaging the sample. If botanical evidence is allowed to freeze, the plant cells could become damaged and then quickly disintegrate, thereby destroying important identification characteristics. Any plant evidence in plastic bags should be placed in a refrigerator as soon as is practicable and examined quickly thereafter. Refrigeration of fresh plant samples will slow decomposition and deterioration of the sample, but it will not stop or completely prevent deterioration (Figure 3.3). Therefore any delay before the examination of refrigerated material will result in further degradation of the material, making drawing conclusions much more difficult. The only method to properly preserve plant evidence is to correctly dry the material.

The crime scene investigator can assist the eventual botanical evaluation by sampling the plants immediately surrounding the body or crime scene, and noting the information listed in Appendix 3.1. The same information should be recorded for all plant samples. When possible, smaller plants should be collected in their entirety. The whole plant, including roots, should be removed from the ground with a hand trowel or shovel (Figure 3.4). Once removed, the soil should be gently separated from the root system and the plant specimen placed in paper. If the plant is too large for the paper or container it can be folded, zigzag or accordion style (Figure 3.5). If the plant cannot be folded, it can be cut in half or into sections (Figure 3.6). Each cut section can be put into separate paper and all sections of the same plant should have the same evidence number. Each section should be marked as top, middle, or bottom.

Figure 3.4   Small plants should be removed from the ground with a trowel or small shovel so their root system is preserved. (Courtesy of Dr. J. H. Byrd)

Figure 3.5   (a) Larger plants can be cut to ground level with a knife or pruning shears. (b) Many plants will have stems that are too long for a botanical press or evidence packaging. (c) These plants can be folded into an accordion or zig-zag pattern for transport and preservation. (d) Once folded, even dried specimens can be easily unfolded for botanical examination. (Courtesy of Dr. J. H. Byrd)

Figure 3.5 (*Continued*)

For extremely large plants, only a portion of the plant needs to be collected (Figure 3.7). For instance a 20–30 cm portion of a branch or vine, with its accompanying leaves, should be satisfactory for identification. A single species of plant can be as variable as individual humans, therefore several samples of the same plant can be extremely helpful in determining the variation that exists. Be certain to look for flowers and fruits, and collect several samples of each. These pieces can be trimmed with sharp pruners or a sharp knife. Dull tools can cause significant damage to plant samples and their use should be avoided.

Very small plant fragments or seeds can be put into a packet made from a folded piece of paper, called a druggist fold (Figure 3.8). Alternatively, thick roots, branches, stems, and bark can sometimes be split so that they conveniently fit into collection paper or boxes. It is very important to look under and around the collected plant for fallen leaves, fruits, flowers, or other parts that may be valuable to the investigation. If several of the same kinds of plants are in the area, always try to collect a plant with fruits or flowers.

Figure 3.6   (a) Plants with large leaves may be encountered as physical evidence and these can be preserved in a standard plant press. (b) Many plant leaves will be too large to preserve as a whole specimen. (c) Large leaves can be cut in half so that one side of symmetry is preserved. (d) Once cut in half the leaf can be folded as shown in Figure 3.5d. (Courtesy of Dr. J. H. Byrd)

All plants (even if they are the same kind) do not necessarily flower at the same time. By looking around the scene for flowers or fruits on identical plants you will greatly enhance the chances of identification.

Succulent plants, or samples with watery or juicy fruits such as an orange, should either be sliced thinly enough to place into newsprint or placed whole into a pasteboard box. Since sugar-filled fruits and excessively wet samples will stick to newsprint, such samples should have waxed paper placed on both sides or around them. This will keep the sample from sticking to the paper or cardboard container and still permit proper drying. If plant material becomes stuck to the paper or cardboard container, determination and analysis may be impossible.

Many, if not all, outdoor crime scenes have plants present. If not present directly within the scene, they are often in the immediate vicinity. Additionally, many homes contain flower arrangements and/or potted plants. Most commercial buildings have ornamental plants outside, including lawns and lawn weeds. Potted plants, flower arrangements, and landscaping may be damaged or in disarray after a crime. The suspect may inadvertently walk across lawns, indicating the path of ingress or egress. At any crime scene, damaged plants should be photographed and collected as a matter of routine standard operating

(c)

(d)

Figure 3.6 *(Continued)*

procedure (Figure 3.9). Likewise, weeds that are common and often overlooked in nature are also frequently ignored in forensic science. Weeds can be extremely valuable in an investigation because of their adaptations for seed transport. Also, the investigator should not forget that pollen can be everywhere.

If not preserved flat and supported by a rigid surface, a dried plant can be easily crushed when any heavy object is placed on top of it. A plant press is the best device to

Figure 3.7 For extremely large plants such as shrubs and trees, a 20–30 cm portion of a branch is all that is necessary to preserve once the plant is properly photographed as part of the crime scene documentation. (Courtesy of Dr. J. H. Byrd)

preserve and flatten plant evidence and facilitate the drying process. A plant press has a hard wooden board or wood lattice on each side and is filled with successive layers of cardboard and newsprint (containing the sample), with blotter paper on each side of the newsprint (Figure 3.10). The blotters on each side of the newsprint will aid in removing the moisture and helping to dry the plant so that it will not deteriorate. The corrugated cardboard also helps to remove moisture. Cardboard absorbs moisture and channels it out through the corrugations. Dry air can be channeled through the cardboard to help quickly remove moisture.

Figure 3.8 A "druggist fold" commonly utilized to package and store small items of physical evidence. (Courtesy of Amanda Fitch. M.S.)

Figure 3.9 All plants present at a crime scene should be inspected for damage that may have occurred during the commission of the crime. If damage is found, written notations, photographs, and plant collections should be made. (Courtesy of Dr. J. H. Byrd)

Given that field conditions can be difficult, a handy plant collection device, called a field press, can easily be made using cardboard. Two pieces of cardboard are cut to the size of a once-folded page of newspaper. Newspaper can then be folded to this size (newspapers are frequently delivered folded only once), and when all samples are in newspaper they should be stacked. One piece of the cardboard is placed on each side of the stacked samples. Any rope, webbing, belting, etc. can be used to secure the stack

Figure 3.10 (a) The plant press functions by keeping the dried plant material rigidly supported and compressed between the alternating layers of cardboard and blotter paper. (b) Each plant specimen should be placed between sheets of blotter paper, held between sheets of corrugated cardboard, and pressed with the wooden framework of the press. This is held in place with two straps. (Courtesy of Dr. J. H. Byrd)

Figure 3.10   (*Continued*)

(Figure 3.11). This bundle is ready for transport and easily carried. If a field press cannot be made, a phone book or magazine can be utilized. Each page containing a plant sample should be carefully numbered just as the individual sheets or newspapers are in a standard press. Phone books are useful because they have very absorbent paper and most are heavy, thereby greatly aiding the drying process. Some magazines use slick or wax-coated paper stock. If using a magazine with slick paper, the samples should be placed into newsprint as soon as possible.

Figure 3.11   A completed field press, which can be readily made and used when a botanical press is not available. (Courtesy of Dr. J. H. Byrd)

Collected plant evidence should be taken immediately to a botanist for separation, drying, and analysis. Involving a botanist shortly after the collection process is completed will ensure that any deterioration of the sample from improper preservation is minimized. If the investigator is not going to deliver the plant material to a botanist immediately, the material in the plant press will need to be separated to ensure proper drying. When separated, each sheet with a sample should be placed under a weight (such as a large book) to make sure it will be flat when dried. Using cardboard for separators, sheets with plants can be placed in a stack with a heavy weight (such as a large book) on top. Drying a normal, not fleshy or very wet, sample will take 2 to 7 days in an air-conditioned room. Drying can be faster if using a drying oven or other specialized equipment. If using an oven, the heat should be set at the lowest or "warm" setting. Most plant identification experts will have laboratory equipment to aid drying. Any collected fungi samples should be left in a box to slowly air dry.

Initial crime scene notation

When doing the initial walk-through of a crime scene containing botanical evidence, there should be a review to determine if the plant material present may show how the suspect, victim, or vehicle entered and exited the area (Figure 3.12). The walk-through should include documentation of mashed or broken branches, or any vegetation that has been pushed aside. Walking, physically carrying a body, or the use of a vehicle will cause damage to vegetation around or leading into the scene. Following a trail of damage to plants can expand the parameters of a scene, which may provide valuable information for the investigation. Photographs of the disturbance and samples of each disturbed plant should be taken and the plants collected prior to the collection of any other type of physical evidence (Figure 3.13).

Figure 3.12 Dr David W. Hall conducting a scene walk-through documenting the damaged plants at a scene which will need to be collected by the crime scene analyst. (Courtesy of Gary D. Knight)

Figure 3.13   Each item of plant damage at a scene should be photographed extensively (both with and without measurement scales) before being collected. (Courtesy of Dr. J. L. Castner)

It is important to remember that all plants at a crime scene, indoors or outdoors, may be valuable evidence, therefore the entire area should be carefully photographed before any evidence is processed. Photographs of the scene should show at least each cardinal direction and include the suspected route of access to the site. In addition to photography, videography can be useful as a supplemental tool, but should never be used in place of still photography. The most common problem in reviewing videotape footage of a crime scene is blurred images. When videotaping, the movement of the camera should be extremely slow. If done slowly the process will limit sudden jerks and allow the viewer to freeze the video to carefully review objects.

In general, the method of collection should be as follows: (i) collect any evidence of the suspect's pathway leading to and from the scene, (ii) collect evidence from the perimeter, especially plants typical of the surrounding habitat, and (iii) carefully search to find any remaining evidence throughout the scene.

Where to search for evidence

Clothing and hair are obvious places to search for evidence, but everything exposed to plant material should be carefully examined, including pockets, cuffs, and textile seams, which can easily conceal evidence. Other frequently overlooked items are shoelaces and shoe seams (Figure 3.14). These are places which often contain small traces of botanical evidence. When dealing with vehicles as evidence, always search the undercarriage, chassis, and wheel wells (Figure 3.15). The front of the vehicle should not be overlooked. Grooves on and around the front bumper and radiator should be examined for plant material. Small plants and fragments can often stick to wet smooth surfaces and can be found on any portion of an automobile.

Natural glandular secretions from leaves, fruits, seeds, and other plant parts serve to help plants naturally move from place to place, and these characteristics provide

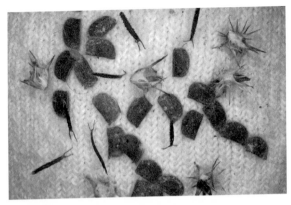

Figure 3.14 Plant seeds that have become attached to a sock provided valuable evidence to link the suspect to the crime scene. (Courtesy of Dr. D. W. Hall)

a means of attachment to many surfaces. Bruised or broken plants can also leak sap, which can enable plant parts to stick to various surfaces. The spines and thorns found on many plants can pierce skin, clothing, and other substances. Also, there are dozens of kinds of hairs found on plants, and these enable plant pieces to adhere to clothing and other surfaces. Most suspects overlook plant evidence because they have no knowledge of its possible importance. If warranted, to reduce the chances of missed plant evidence, have a botanist recheck all clothing and other evidence retrieved from an outdoor scene because trained eyes are more likely to see small pieces of plants.

Perennial plants such as trees, which grow for several to many years, often contain seasonal and/or annual rings. If a plant with annual rings is growing over, through, or closely adjacent to a body, skeletal remains, a gravesite, or object at a crime scene, the rings can be counted and provide a time interval. Often, a time interval can be established for damage to trees and large branches such as could be caused by a vehicle running into a tree, a vehicle parked on the roots, soil being placed over a portion of the roots, flooding, shading of the foliage, unusual temperatures, and other disturbances. Other perennial and even annual species can be used similarly in certain situations.

To collect a tree in order to preserve the rings or other significant damage, the tree and any damage to it should be carefully photographed. If the tree is too large to collect in its entirety, the trunk should be cut off a few inches below and above any damage. This section should be kept in a cardboard box in the same way as for any other larger plant evidence. Be sure to collect samples of the leaves and any flowers or fruits to aid the identification of the tree. Samples of the leaves or other vegetation should be preserved separately, but tied together by the same item/sample number. The photographs and evidence for all of these samples must be linked in order to maintain the legal requirements for physical evidence. An easy way to do this is to record the item number within the photograph and list the item number in the evidence log and written notations.

Figure 3.15 Plant fragments that have become attached to the undercarriage of an automobile provided valuable information which linked the vehicle to the scene. (Courtesy of Dr. J. H. Byrd)

Smaller broken branches should be carefully photographed, especially the ends within the broken area, any evident sap, and the branch tip showing the leaf wilt, if any. The branch should be cut at least 15 cm below the break and the entire branch collected. The sample should be placed in newspaper, a bag, or a box. Be sure to collect any flowers or fruits on the plant. Preserve leafy material flat.

Any plant part touching or buried with human remains can be valuable evidence (Figure 3.16). Frequently the time of death, time of year, or prior locations can be indicated from plant material. If a botanist is not on site, color photographs and preservation of all associated plant material are crucial. Footprints on plant material, branches which have been broken, as well as all the vegetative material directly under, on, or buried with the remains should be collected. Collection of evidence concerning time of death is best conducted at the scene by a trained botanist. However, examination of high-quality crime scene photographs can be a useful tool in estimating the time of death if a visit to the scene by a botanist is not possible.

When someone digs a hole to bury remains they usually disturb and damage the associated plants to some extent. Soil from the hole is usually returned to cover the

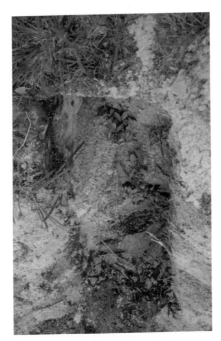

Figure 3.16 Botanical materials recovered from clandestine gravesites should be thoroughly documented and submitted to a forensic botanist for analysis. (Courtesy of Dr. J. H. Byrd)

remains. Subsequent raking or shoveling the excavated soil into the hole will damage additional plants around the burial. Attention should especially be directed to the area where the soil from the hole was piled. Photograph all disturbed areas around the hole carefully (Figure 3.17).

Figure 3.17 Plants growing in the soil over the gravesite and in the disturbed soil around the grave site should be documented and collected for analysis by a forensic botanist. (Courtesy of Dr. J. H. Byrd)

Figure 3.18 Cut roots define the edge of the original burial pit in this excavation of a clandestine gravesite. (Courtesy of Dr. J. H. Byrd) (Please refer to the colour plate section.)

Plant fragments, sometimes numbering in the thousands, can be buried with remains. All fragments should be saved, although this can be time-consuming and greatly slow the progress of remains recovery. Screening is usually the best means of finding small fragments. Photographs of each level of recovery will document that the plant material was actually buried with the remains.

Roots can be equally valuable. The roots of some species of plants produce annual rings, but even without annual rings roots can still be used to show relative time intervals. Roots that have been cut can indicate the edges of a burial pit (Figure 3.18). When a root is cut it may branch out and continue to grow after it is reburied. Preserving this new growth intact may enable the determination of the approximate time when the root was originally cut.

Roots should be photographed when encountered. If roots are to be collected, they should be pruned some distance from the damaged area to preserve the internal structure at the point of contact. Many species of plants are difficult to identify from roots alone, even for a specialist, as roots of many species appear quite similar. At the scene it is useful to try to follow the roots to their source at the plant and preserve a sample of that plant to assist with identification. Roots frequently intermingle or grow for considerable distances, thus following the root to its source is often a difficult and time-consuming task. Roots are collected as any other plant evidence: if small they can be placed between paper and if larger they can be put into a bag or box. The etching of bone by roots is a common occurrence in cases of extended time period burials. The depth of etching may be helpful in determining time intervals, but no research has been produced which shows the time interval necessary for the roots of various species of plants to etch bones.

Storage

Plant evidence should be stored in a cool dry area. Plants are biological organisms and need to be killed and dried for the best preservation. On reaching the storage or processing area, each sample should be separated until dry. Drying will kill the plant and stop its biological processes. Drying is best if the plants are kept flat with cardboard and/or blotter paper on both sides and a weight placed on top. Several plants can be piled on top of each other as long as they are kept separated by cardboard. A commercial plant press can simply be left to dry as collected. Fungi and other large plants or plant parts collected in cardboard boxes should be left in the boxes. Most plants kept in this manner will dry in 3–7 days if humidity is low. Wet plants will need to have the surrounding paper replaced each day for best results. Special plant dryers using very low heat can be purchased or constructed. Once dry, if not kept in the phone directory or magazine, each newsprint or bag with a sample should be removed from the cardboard and/or blotters and put into a cardboard box. The samples should be kept in numerical order for ease of processing. If the box is large enough, other samples in smaller boxes or bags can be included. Phone directories or magazines with samples should also be placed in a cardboard box. The box should have a few mothballs or a no-pest strip included to repel any insects that may eat dried plants.

Documentation of botanical evidence

For plant material to be valuable as evidence, it is critical to record several types of information as a crime scene is processed. A sample data sheet is given in Appendix 3.2.

Environmental conditions are frequently overlooked and not recorded, but should always be noted: is the weather sunny/clear/partly cloudy/rainy (how much rain), is there a slight breeze, is it breezy or windy, what is the humidity, and what is the temperature?

In addition to general environmental conditions, other relevant information about scene conditions should also be recorded. Descriptions of the types of information that should be documented, preferably on a botany field data sheet, are as follows:

- *Habitat*: The particular habitat (place) at the scene. The scene may be in a forest, but the sample may have been in an opening, ditch, or trail. Without knowing the habitat, much of the plant evidence cannot be interpreted. Is the site forested, shrubby, an open field, a roadside ditch, a home lawn? Is it a landscape planting, an industrial site, etc.?

- *Scene location*: Usually only one scene location and description is needed and can be listed at the beginning of a log sheet. The scene must be linked to a permanent structure, such as a corner of a building, a utility pole (poles usually have an identification number), or a corner of a roadway. Local building codes require that permanent structures have surveyed locations so that records will indicate the exact location of the scene, even long after a structure has been destroyed. A physical description of the location should always be made, although easily available geographic positioning systems (GPS) are common and can be used to provide a second confirmation of the location.

- *Evidence location*: Where was this evidence collected at the scene: on the body or vehicle, or elsewhere? Where on the body or vehicle was it found? If not on a body or vehicle, the evidence will require exact measurements for its location.

- *Collector*: Who actually collected the evidence? It is best if only one individual collects all of the plant evidence as chain of custody is vitally important. Each person who handles any evidence *must* be documented by a chain of evidence log.

- *Agency*: The agency and title of the plant collector must be noted.

- *Date*: The date of collection must be included in the log.

- *Item number*: Each plant sample must have a unique number.

- *Time of collection*: Note the time when each sample was collected.

- *Type of plant*: (if known): tree, shrub, herbaceous.

- *Height of plant*: (if known, and the entire plant is not collected): This measurement does not need to be exact, simply an estimate.

- *Flower color*: Color is often subject to change on drying.

- *Fruit color*: Color is often subject to change on drying.

- *Fruit shape*: Shape is subject to change on drying.

- *Frequency*: Look around to determine how common your plant is: common, frequent, infrequent, occasional, rare (only one seen).

How to have botanical evidence analysed

There are distinct advantages to working with a local botanist rather than one who is unfamiliar with local habitats. A local botanist who thoroughly knows the plants in an area is the best person to work with the evidence. While there are approximately 350,000 to 400,000 different kinds of plants in the world, often fewer than 100 will occur in the immediate area of a crime scene. Someone who knows the local plants is more likely to be able to identify the evidence quickly, often at sight. Plants throughout the world are identified by their flowers and fruits, but flowers and fruits may not be present on the evidence. Local botanists are frequently very good at identification of the plants in their area using only vegetation, even fragments of vegetation, but a botanist from a different area may not be able to identify the plants easily, if at all. To avoid wasting time and money, finding a local botanist is a good idea. If the local botanist is reluctant or has problems with procedures for case analysis, help can be quickly obtained by a phone call to an experienced colleague.

Every agency or company that deals with plant evidence should have a botanist on-call so that delays in analysis are minimal. A call in the middle of the night is startling in itself and the invitation to a crime scene can be shocking. The local botanist can become familiar with the parameters of the collection of evidence at a crime scene, help the agency assemble the materials that will be needed for collection, and make preliminary contacts with other botanists who might provide additional information. Most botanists are uninformed about the forensic sciences in general and, if available to help, will require lead time to understand the discipline and the chain of custody requirements. Every forensic botanist will need to set up a system of other experts who may be called upon to help. Such a network takes time to organize. Each expert in the network may require time to understand the forensic implications of his or her discipline. A good place to start is with a local botanist who teaches or is responsible for plant identi-fication. This person should be able to suggest other experts when needed.

Where to find a botanist

Virtually all community colleges, colleges, and universities have a botanist on their faculty. Other than academic institutions, professionals with botanical training and knowledge can be found in the United States at county extension offices and worldwide at forestry companies, parks, horticultural businesses, and various government offices.

Types of cases

Many botanists who teach at any level of college and those who work for government agencies have done some type of forensic work, but they may not be aware of it. Since forensics involves the law, some forensic situations simply involve the determination of a plant prohibited by law (Figure 3.19). Laws list some plants used as medicines or drugs or those believed to be noxious weeds as illegal. Therefore possession of any of these listed plants is illegal, but the first step is identification, which often demands someone with special training. Although involving different areas of law, it is often as important for society to get rid of a major pest plant as it is to place a suspect at the scene of a crime or to determine the time since death. Major pest plants can cause millions of dollars in damages. Certain pest plants can change our entire environment, cause loss of crops, leading to high economic losses or even starvation, be toxic (external or internal) to sensitive individuals, or be the physical cause of death (punctures, weapons).

Evidence analysis

The initial review of plant evidence should provide an analysis of the possibilities as well as the limitations of the material. Time and costs for further analysis should be part of the review. Evidence analysis can require special equipment such as a scanning electron microscope or other optical scopes (see Chapter 6), chemicals, trips to a scene,

Figure 3.19   Marijuana is an illegal drug plant in many countries. (Courtesy of Dr. D. W. Hall)

special collection techniques, growing plants, unusual preparation for analysis, or additional information on different methodologies.

The botanist should provide a realistic estimation of the time needed to evaluate the evidence, and provide a written result and, at least, an oral report. If the evaluation takes too long, the necessity for the analysis may not be needed, for example the time limit for prosecution may run past the deadline. Since plants are biological organisms they can continue to grow, making evidence at a scene, if delayed for too long, irrelevant. Plant evidence at scenes will continue to change. Temperatures, seasons, angle of sunlight, ground cover, seed/propagule germination, recovery after damage, and other factors can all be relevant and variable. Although a visit to the scene is always best, some evaluation can be provided from photographs, but this is subject to problems of interpretation. Photographs must be of good to excellent quality and show the relevant habitats, plants, and plant parts (Figure 3.20). Unfortunately, photographers are often not aware of the types of photographs necessary for botanical analysis. A timely call to a forewarned botanist can be invaluable for the photographer and may avoid a visit to the scene by the botanist.

Costs which translate into time and materials enter into all evaluations. The common types of forensic plant evidence can be evaluated very quickly, often in less than an hour, so that costs are minimal or not necessary. However, if not preserved correctly, evidence that requires special techniques can deteriorate and become totally useless. Evidence that will require cellular analysis or unusual identification procedures can be very time-consuming. The cost may exceed a budget, although many scientists employed by public institutions will provide an analysis for no cost or for expenses. No botanical specialists, other than those involved with DNA, are known to be employed by publicly funded forensic laboratories.

To avoid misunderstandings, a written contract should be established between the parties, even for examinations where no fees or expenses are required; often an email or fax can suffice.

Figure 3.20 (a) Photographs of plants taken at a distance, or of plants in groups, may not show the detail a botanist requires when making an identification. (b) A detailed photograph of a leaf or terminal end of a stem is the type of detail that should be recorded for the botanist. (Courtesy of Dr. J. H. Byrd)

Laboratory report

A laboratory examination form should be compiled for each container (Appendix 3.3). The form should list the date, case number, agency or company, contact person, method of contact, method of transport, and type of packaging. The form should document the kind of container, the date opened, and how the evidence was contained after opening. If several samples are in the same container they can be described separately on the same form. Care must be taken that each sample has a separate number. If unnumbered items are separated from a larger group, the examiner should place a unique sample number on each item and package them accordingly.

The method of analysis of each item should be detailed. This can be as simple as examination by eye or hand lens, or as complex as any machine might make it. If a machine is used for analysis, a procedures manual should be kept handy as a reference during the examination and for any testimony afterwards. A step-by-step analysis should be detailed, but if the analysis is standard a manual can be listed instead.

The results of each analysis must be noted, whether positive or negative, with as much information as can be discerned. If a leaf or stem fragment is examined and a species cannot be determined, a simple note that the fragment is from a dicotole-dyonous (dicot) plant or an unknown woody plant might suffice. If the client wishes to pursue additional tests, other than those previously authorized, the methods and costs must be discussed with the client and a written authorization (a contract amendment) should be required.

A written report should only be prepared if requested by the client. The report must include a statement of how the evidence applies to the case. The examiner may need to ask for certain particulars of the case to provide a relevant analysis. Scientists should refrain from obtaining too many details to avoid prejudice.

Transportation of botanical evidence

The chain of custody is critical for most forensic cases, especially so for those criminal in nature. The chain of custody ensures that the evidence is secured and in someone's possession or control at all times. A log sheet for possession of the evidence is kept with the evidence (see Appendix 3.4). The person who collects the evidence must keep the evidence in his or her possession in a secure place, usually somewhere with a lock. If the evidence is given to someone else it must be someone who is authorized to accept it, such as the keeper of a locked evidence storage unit or the director of a laboratory. The person who accepts the evidence must sign, date, and note the time on the log sheet. The evidence is presumed to be in control of that person upon transfer.

The transportation of the evidence can be by any methods that will ensure control and maintain chain of custody. The person holding the evidence may transport it themselves or by other carriers, such as Federal Express, a government mail service, courier, or any law enforcement personnel. Very sensitive evidence may require the courier to maintain the evidence in their possession at all times. A courier may stay with the evidence while it is analysed.

Evidence retention and disposition

The evidence can be retained by the examiner or returned to the law enforcement unit or attorney. Before the evidence leaves the examiner's possession, it should be resealed with tamper-evident tape. The examiner's initials or signature together with

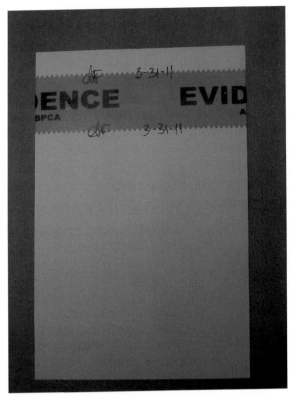

Figure 3.21   A paper evidence bag illustrating proper labeling methods with signature or initials on both the packaging and tamper-evident tape. (Courtesy of Amanda Fitch, M.S.)

the date should be written across the evidence tape so that a portion of the signature is also on the packaging (Figure 3.21). Evidence tape is commonly available from police and forensic science equipment suppliers. If the evidence is transferred and retained by the examiner, it must be kept secured until authorization is given to destroy it.

All evidence that is not destroyed by testing should be returned on request by the client. The evidence might include prepared samples. If the original container is destroyed, a clearly labeled replacement container should be used. All remains of the original container should be included in the replacement container. Any evidence kept by the examiner should be kept in a secure place, such as a locked cabinet or closet.

When evidence is returned, a record of the method used for return together with any receipts and the time the evidence left the examiner's control should be noted on the examination form. The client/agency should be consulted to determine the method of return, and written documentation always obtained for the return from the client/agency.

Step-wise method for the collection of botanical evidence

1. Set up a field or botanical press.

   a. Newsprint or other cheap absorbent paper should be folded over each sample.

   b. Each piece of evidence should have a case number and a distinct evidence number. Any writing, including numbering, should be done with a pencil or indelible marker containing ink that will not run or blot under moist conditions. Many types of ink will run or blot and can become indecipherable if they become wet.

   c. Use wax paper or shiny photographic quality paper to collect wet samples, such as those in watery or fleshy fruits, so that they will not stick to the paper while drying.

   d. Paper grocery-type bags and cardboard boxes can be used for larger samples or for cones or nuts that will not press flat.

2. Collection of plant material around and in the scene should include (see photographic examples) at least 10–12 samples of different plants around the scene to document local vegetation.

   a. Prune 20–30 cm samples of branches of woody plants.

   b. Dig up herbaceous plants and include roots.

   c. If the sample is longer than the press or box, fold each specimen in an accordion zigzag-like manner.

   d. If evidence is put into a paper bag, if possible the bag should be mashed and kept flat, just as for a sample put into a press. If a press is available, the flattened bag can be placed under a flat weight such as a large book. The flattened plant is more easily stored and less likely to incur damage by crushing of the dried material.

3. Each item of evidence should be photographed before it is collected and photographed again as it is placed into the press, bag, or box. The photograph of the plant being placed into the press, bag, or box should show the evidence number adjacent to the sample.

Equipment required

The basic equipment needed for the collection of botanical evidence is commonly available at most general stores. Some of the basic items needs are hand pruners, garden spade, archaeological (or masons) trowel, evidence numbering flags or tents, digital SLR camera, graphite pencils (at least two), wax paper (or alternatively magazines with coated photographic paper), paper grocery bags, cardboard boxes (various sizes), plant press, tamper-evident tape, and evidence and chain of custody forms.

Plant Presses

1. A formal press can be obtained from most biological supply houses. It is composed of a stiff board or other support on each side with alternating: cardboard, blotter, newsprint, blotter, cardboard, blotter, newsprint, blotter, cardboard, etc. The entire press is tightly bound with straps to keep the evidence flat. The covers, cardboard, blotters, and straps can be purchased separately.

2. A field press can be constructed with simple materials available by cutting two pieces approximately 20–30 cm in size from a cardboard box, then simply using newspaper to hold each specimen, and stacking the newspapers with specimens on top of each other. A strap, piece of tape, string, or belt can be used to bind the cardboards and specimens for transportation.

3. Other types of field presses can be simply fashioned using of a phone directory or magazine and putting samples between the pages.

Appendix 3.1

Crime scene data

For plant material to be valuable as evidence, it is critical to record several types of information as a crime scene is processed. A sample data sheet is given in Appendix 3.2 to assist the investigator in the organization and documentation of collected plant material. Environmental conditions are frequently overlooked and often not recorded contemporaneously while documenting the scene. Basic weather conditions should be noted, such as sunlight, rain, clouds, wind, temperature. and humidity.

Habitat documentation

Without knowing the habitat of the crime scene, much of the collected plant evidence cannot be interpreted. Basic habitat documentation should include information on the site, i.e. forested, open field, roadside ditch, lawn, or industrial site.

Scene location

Usually only one scene location and description is needed and can be listed at the beginning of a log sheet. The scene must be linked to a permanent structure, called a datum, such as a corner of a building, a utility pole (poles usually have an identification number), or a corner of a documented roadway. Local building codes require that permanent structures have surveyed locations so that records will indicate the exact location of the scene, even long after a structure has been destroyed. A physical description of the location should always be made, although easily available geographic positioning systems (GPS) are common and can be used to provide a second confirmation of the location.

Collection information needed for each botanical sample

- *Evidence location*: This should include information on exactly where at the crime scene was the evidence located, such as on the body, vehicle, or elsewhere and where on each was it found. If not on a body or vehicle, the evidence will require exact measurements for its location based on a coordinate system with the datum as a point of reference.

- *Collector*: All collected botanical evidence should include documentation of the individual who collected the item, therefore it is usually best if only one individual collects all of the plant evidence at a scene as this makes for easier documentation on the chain of custody evidence log.

- *Agency*: The agency name and title of the plant collector should be noted.

- *Date/time*: The date and time of collection should be included on the evidence container and noted on the evidence log.

- *Item number*: Each plant sample should have a unique item number to eliminate possible confusion with any other sample. If an individually numbered tamper-evident bag or seal is used, documentation of the number must be made in the scene notes.

- *Type of plant*: The general type of plant should be noted (tree, shrub, herbaceous).

- *Height of plant*: If the entire plant is not collected a height estimate should be made.

- *Flower color*: Flower color often changes on drying so the original color should be documented with photography. If possible a photo-grey scale (18% grey) should be included in the photograph.

- *Fruit color*: Fruit color often changes on drying so the original color should be documented with photography. If possible a photo-grey scale (18% grey) should be included in the photograph.

- *Fruit shape*: Shape is subject to change on drying so the original shape should be documented with photography.

- *Habitat*: The particular habitat found at the scene should be documented photographically and with written notations.

- *Plant frequency*: The relative abundance of plant species at a crime scene should be documented both photographically and in writing. Make notes on whether a particular species is common, frequent, infrequent, occasional, or rare.

Appendix 3.2

Botany field data sheet

Case number:_____Date:_____

Agency:_____

Investigator: _____

Scene location: _____

Description of the scene:_____

_____

_____

_____

_____

Description of biological conditions at scene:

_____

_____

_____

_____

Habitat assessment
Description of habitat
Describe:
    Obvious path, trail, or road to/through scene:

    Broken branches:

    Other obvious habitat disturbance(s):

Habitat sampling procedure
Include list of trees, shrubs, ground cover, and herbaceous plants

Item #	Description	Location	Disposition

Visible botanical evidence description
Describe:
  Plant material covering body, vehicle, or other object:

  Other plant evidence on body, vehicle or other object:

  Plant evidence under or buried with body, vehicle, or other object:

Sampling procedure for botanical evidence
Include list of botanical evidence collected

Item #	Description	Location	Disposition

Evidence transportation

From	Name/signature		Date	
To	Name/signature		Date	
From	Name/signature		Date	
To	Name/signature		Date	
From	Name/signature		Date	
To	Name/signature		Date	

Evidence storage
Include list of items of evidence being stored

Item #	Description	Location	Disposition

Evidence analysis
Name of botanical analyst:

Location of analysis:

Conclusions

Appendix 3.3

Botany laboratory examination data format

Date: _____

Case number:_____

Agency (or company): _____

Contact person:_____

Method of contact (phone, personal visit, etc.):_____

Method of transport (FedEx, US Mail, in person, etc.):_____

Type of packaging (box, paper bags, truck load, plastic bag, etc.):

_____

Information documented for each container:

1. What kind of container?
2. Date opened.
3. How is evidence contained after opening?
4. Contents of container.

    a. Number of samples.

    b. ID of samples or best description of each item if not able to ID.

5. Number of each identified (or unidentified) item per container.
6. Analysis.

    a. Method of analysis (dissecting scope, microscope, hand lens, SEM (include lens type and magnification).

    b. Preparation details, if sample needs to be cut, stained, or otherwise prepared for analysis – should be step by step (can use standard steps from manuals if they directly apply to methodology).

    c. If written report is requested, include statement of how this evidence applies to case.

    • Only prepare written analysis if requested by client/agency

    • Ask for certain particulars of case to provide relevant analysis (scientist should refrain from getting too many details to avoid prejudice).

7. Return all evidence (including any prepared samples to original container or into clearly labeled replacement container if original container destroyed or if samples not presented in a container).

    a. If using a replacement container, save any remnants of original container and place into new container with evidence.

8. If evidence kept by investigator, all evidence should be in a secure place (locked case, cabinet, closet, room).

9. If or when evidence returned, keep record of time and method of return.

    a. Consult client/agency to determine the method and location of the return, and the individual and agency who will receive it.

    b. Obtain written documentation of return from the client/agency.

Appendix 3.4

Evidence log

Agency name: Case number:

Scene location:

City, state:

Item #	Description	Location	Disposition

Item #	Description	Location	Disposition

4 Expert evidence

Bernard A. Raum JD, MFS

"[t]he whole object of the expert is to tell the jury, not facts,... but general truths derived from his specialized experience."

Judge, Learned Hand[1]

The common law

The evidentiary rules relating to the receipt of expert evidence in both the United States and the United Kingdom have the same early roots in English Common Law. There exists a history, as early as Bracton of the impaneling of "special" juries composed of persons who possessed exceptional knowledge specific to the facts at issue,[2] i.e., "experts". This practice apparently continued into the 19th century in England. Of course, on these occasions the "experts" were acting as the trier of fact rather than as expert witnesses *per se*. On a parallel path, the law regarding "opinion" testimony by "expert" witnesses was slowly developed, seemingly on an *ad hoc* basis.[3] In a series of cases during the 18th century, for example, physicians were called to testify concerning the cause of death and, without discussion, were allowed to testify. In the later part of the 18th century, in a series of judicial decisions beginning, most notably, with the decision by Lord Mansfield in *Folkes v. Chadd*,[4] the English court focused specifically on the question of the admissibility of expert "opinion" testimony.

Originally, the so-called "opinion rule" barred opinion testimony by witnesses in general on the theory that it was "mere opinion" rather than an offering of facts based on the personal knowledge and observations of the witnesses;[5] without personal

---

[1] Learned Hand, *Historical and Practical Consideratons Regarding Expert Testimony*, 15 *Harv. Law Rev.* 40 (1901).

[2] Under the Writ *de venire inspeciciendo*; see Learned Hand, *supra*, p. 44 and the cases cited *passim* beginning in the year 1343; also see Hodgkinson, T. and James, M., *Expert Evidence: Law & Practice*, 3rd edn (2010), §1-009, p. 8.

[3] *Ibid.*

[4] (1782) 3 Doug. 157.

[5] *Carter v. Boehm* (1766) 3 Burr. 1905, and see in general Wigmore, J. H., *A Treatise on the System of Evidence in Trials at Common Law*, 2nd edn (1904), § 1917, and the discussion in Hodgkinson, *supra*, Chapter 1.

Forensic Botany: A Practical Guide, First Edition. David W. Hall and Jason H. Byrd.
© 2012 John Wiley & Sons, Ltd. Published 2012 by John Wiley & Sons, Ltd.

knowledge of the facts, the witness was regarded as incompetent to testify. At this same time, an apparent exception to this rule was recognized concerning the testimony and evidence of skilled witnesses.[6] This was not actually a new concept but rather was one well known to the Common Law courts as early as the 14th century, where such skilled witnesses were commonly looked to for advice by the court rather than as an aid to a jury.[7] It was only in the 18th century that this skilled person came under the scope of the "opinion rule".[8]

Another consideration in the evolution of expert evidence is the concept that in order to be admissible, the expert opinion must actually aid the jury in resolving a fact in issue,[9] subject to the caveat that the jury must make the ultimate findings of fact. This, of course, became a lynch-pin test for the admissibility of such evidence[10] *which is still in full force in both the United States and the United Kingdom.* During this ongoing evolution, it also became generally recognized that an expert, in rendering an opinion, was no longer required to base their opinion on facts within their personal knowledge or personal observations.

The United States experience

For the balance of this discussion, it is well to keep in mind that the initial decision to admit the testimony of *any* witness rested (and still rests under FRE 104) within the exercise of the sound discretion of the trial judge and would not be overruled on appeal unless that discretion was "abused," something that very rarely occurred.

A review of the case law in the 19th century in the United States discloses that the accuracy of scientific evidence was usually left to the judgment and credibility of the individual expert witness. The difficulty with this proposition was that neither the jury in deciding the case nor the judge in ruling on admissibility had any objective basis to determine the witness's scientific credibility or the accuracy of any technique or methodology involved.[11]

The first real attempt to create some method for the court to gauge the validity of the science itself came in the watershed decision of the United States Court of Appeals for the District of Columbia in *Frye v. United States.*[12]

---

[6] Wigmore, *ibid.*, p. 2542

[7] For example *Buckley v. Thomas* (1554) 1 Plowd. 122, where Staunford, J., stated: "I grant that if matters arise in our law which concern other sciences or faculties, we commonly apply for the aid of that science or faculty which it concerns, which is an honorable and commendable thing in our law.", as quoted in Wigmore, *supra*, §1917, p. 2543.

[8] For example, *Folkes v. Chadd, supra.*

[9] See, for example, *Beckwith v. Sydebotham* (1807) 1 Camp. 116.

[10] Wigmore, *supra*, p. 2546.

[11] See, for example, the comments found in Faigman, D.L., Kaye, D.H., Saks, M.J., and Sanders, J., *Science in the Law: Standards, Statistics, and Research Issues, American Casebook Series* (2002), § 1-2.1, pp. 3–5.

[12] Raum, B.A., A short primer on the admissibility of forensic science evidence in Tennessee: a checklist, *Tennessee Journal of Law and Policy* 6(2), 161, 169 (2010).

The decision in *Frye v. United States*

In 1923, the US Court of Appeals for the District of Columbia announced an objective standard for the determination of the admissibility of expert testimony based on a scientific test.[13] Restating the general law as to the use of expert evidence as stated in the briefs of counsel the court commented that:

> "The rule is that the opinions of experts or skilled witnesses are admissible in evidence in those cases in which the matter of inquiry is such that inexperienced persons are unlikely to prove capable of forming a correct judgment upon it, for the reason that the subject-matter so far partakes of a science, art, or trade as to require a previous habit or experience or study in it, in order to acquire a knowledge of it. When the question involved does not lie within the range of common experience or common knowledge, but requires special experience or special knowledge, then the opinions of witnesses skilled in that particular science, art, or trade to which the question relates are admissible in evidence."

The court then announced its judgment:

> "Just when a scientific principle or discovery crosses the line between the experimental and demonstrable stages is difficult to define. Somewhere in this twilight zone the evidential force of the principle must be recognized, and while courts will go a long way in admitting expert testimony deduced from a well-recognized scientific principle or discovery, *the thing from which the deduction is made must be sufficiently established to have gained general acceptance in the particular field in which it belongs.*"[14] [emphasis added]

The test for admissibility in *Frye* was a simple creation, implemented with no explanation by the court, and strictly involved determining if a consensus of the experts in a given field agreed that the science was valid. Thus, instead of relying upon the word of one expert, now the courts were asked to rely upon the words of a group of experts, again without any meaningful and independent evaluation by the trier of fact.[15]

Over the next four decades, in decisions too numerous to cite, the holding in *Frye* was adopted on a case-by-case basis in both the federal courts and the courts of the various states as the controlling authority governing the admission of evidence by an expert which was based on the application of a scientific principle, procedure, or methodology.[16] To this day, the rule announced in *Frye* is still the controlling authority in three of

---

[13] *Frye v. United States*, 293 F.1013, 1014 (D.C. Cir 1923).

[14] *Frye, supra*, 1014.

[15] Raum, *supra*, 169.

[16] See the discussion in Faigman *et al., supra, n* §§ 1-2.2, 1-2.3, and 1-2.4, pp. 5–10, for an excellent brief discussion of the general *Frye* experience in the United States.

the largest jurisdictions in the United States, California, Florida, and New York, as well as 14 other states.[17] The rest of the states adopted some form of the *Daubert* rational to determine admissibility of expert testimony.[18]

Unfortunately, the reported case law in the United States on the admissibility of the testimony and evidence of expert botanists is not as instructive as one would hope. There is one decision which has an extensive discussion regarding the "expert" testimony of a botanist wherein the court concluded that, although the witness was an "expert", nonetheless his evidence was excluded due to his lack of particularized experience with purple nightshade, the plant in question.[19] There are, however, a number of cases which offer anecdotal information regarding the fact that botanists have been relied upon by the courts to provide expert evidence in the resolution of issues before those courts.[20]

The codified federal rules of evidence

On 2 January 1975, PL. 93-595 was enacted by the US Congress which provided a sweeping change to the Federal Rules of Evidence (FRE). The original Chapter 7 rules

---

[17] They are: Alabama (excepting the issue of DNA evidence, which is specifically controlled by a separate statute that adopted the *Daubert* decision in that regard), Arizona, the District of Columbia, Illinois, Kansas, Maryland, Massachusetts, Minnesota, Missouri, Nevada, New Jersey, North Dakota, Pennsylvania, and Washington; Giannelli, P.C. and Imwinkelried, E.J., *Scientific Evidence*, 4th edn (2007), § 1.16.

[18] See Lustre, A.B., *Post-Daubert Standards for Admissibility of Scientific and Other Expert Evidence in State Courts*, 90 A.L.R. 5th 453.

[19] *State v. Griswold*, 172 Vt. 443, 782 A.2d 1144 (2001), and see *Kemper v. Commissioner of Internal Revenue*, 30 T.C. 546 (US Tax Ct. 1958) drought not cause of death of trees for purposes of "casualty loss" income tax deduction; and see the court's opinion on direct appeal, 269 F.2d 184 (8th Cir 1959), where the court specifically ruled on the admissibility of the expert botanist's testimony.

[20] *US v. Rothberg et al.*, 351 F. Supp. 1115 (D.Ct E.D. NY 1972), *Cannabis sativa L. case; US v. Larkins and Larkins*, 657 F. Supp. 76 (D.Ct W.D. Kentucky1987), wetlands case; *11 Doors and Casings, More or Less, of Dipteryx Panamensis Imported from Nicaragua, Defendant, Thompson and Thompson, Claimants*, 587 F. Supp. 2d 740 (D.Ct E.D. Va 2008), Importation into the US of an endangered species, *Dipteryx Panamensis*, from Nicaragua without a Certificate of Origin; *Missouri v. Morrow*, 535 S.W.2d 539 (Kansas City Dist. Mo. App. 1976), *Cannabis sativa L.* case; *State v. Jolly* (9th App Dist Ohio App. 1975) LEXIS 6629; *Sierra Club et al. v. Martin et al.*, 71 F. Supp. 2d 1268 (D.Ct N.D. Ga 1996), Injunctive relief, timber cutting, endangered plant species, environmental impact, "biologist" as expert, extensive discussion; *Sanders v. Louisiana, Dept. of Natural Resources*, 973 So. 2d 879 (La.App. 3 Cir. 2007), botanist study of vegetable zones to determine high water mark on lake; *Riverhawks; Northwest Rafters Assoc.; Klamath-Siskiyou Wildlands Center v. US Forest Service et al.*, 228 F. Supp. 2d 1173 (D.Ct Or 2002), expert opining that motorboat wakes have a significant adverse impact on riparian vegetation, including loss of critical protective plant cover for salmonids and aquatic animals; *Hirsch et al. v. Maryland Dept. of Natural Resources, Water Resources Admin*, 42 Md. App. 457; 401 A.2d 491 (Md. App 1979); see also the decision of this case on *Certiorari* 288 Md. 95; 416 A.2d 10 (1980), Tidal wetlands preservation; *Kemper v. Commissioner of Internal Revenue*, 30 T.C. 546 (US Tax Ct. 1958), drought not cause of death of trees for purposes of "casualty loss" income tax deduction; and see the court's opinion on direct appeal, 269 F.2d 184 (8th Cir 1959), where the court specifically ruled on the admissibility of the expert botanist's testimony; *Kahrs International, Inc., v. US*, 31 Intl Trade Rep. (BNA) 2096; 2009 Ct. Intl. Trade LEXIS 110 (United States Ct. of Int Trade), law witness allowed to testify as to the identity of certain wood products under FRE 701; *Herriman v. US*, 8 Cl. Ct. 411; 1985 US Cl. Ct. LEXIS 962 (United States Claims Court), high water mark identified by analysis of vegetation; *City of Cedar Rapids v. Marshall*, 203 N.W. 932 (Iowa 1925), high water mark analysis; *Blue Ocean Preservation Society, Sierra Club, and Greenpeace Foundation v. Watkins, Secretary, Department of Energy et al.*, 767 F. Supp. 1518 (D.Ct Hawaii 1991), Environmental Impact Statement (EIS).

relating to opinion and expert testimony were subsequently amended in 2000 to reflect additional issues which arose as a result of the practice under those rules.[21] The amended rules, which are currently in effect, are as follows:

Rule 701. Opinion Testimony by Lay Witnesses. If the witness is not testifying as an expert, the witness's testimony in the form of opinions or inferences is limited to those opinions or inferences which are (a) rationally based on the perception of the witness, and (b) helpful to a clear understanding of the witness's testimony or the determination of a fact in issue, and (c) not based on scientific, technical, or other specialized knowledge within the scope of Rule 702.

Rule 702. Testimony by Experts. If scientific, technical, or other specialized knowledge will assist the trier of fact to understand the evidence or to determine a fact in issue, a witness qualified as an expert by knowledge, skill, experience, training, or education, may testify thereto in the form of an opinion or otherwise, if (1) the testimony is based upon sufficient facts or data, (2) the testimony is the product of reliable principles and methods, and (3) the witness has applied the principles and methods reliably to the facts of the case.

Rule 703. Bases of Opinion Testimony by Experts. The facts or data in the particular case upon which an expert bases an opinion or inference may be those perceived by or made known to the expert at or before the hearing. If of a type reasonably relied upon by experts in the particular field in forming opinions or inferences upon the subject, the facts or data need not be admissible in evidence in order for the opinion or inference to be admitted. Facts or data that are otherwise inadmissible shall not be disclosed to the jury by the proponent of the opinion or inference unless the court determines that their probative value in assisting the jury to evaluate the expert's opinion substantially outweighs their prejudicial effect.[22]

Rule 704. Opinion on Ultimate Issue.

(a) Except as provided in subdivision (b), testimony in the form of an opinion or inference otherwise admissible is not objectionable because it embraces an ultimate issue to be decided by the trier of fact.

(b) No expert witness testifying with respect to the mental state or condition of a defendant in a criminal case may state an opinion or inference as to whether the defendant did or did not have the mental state or condition constituting an

---

[21] The decision in *Daubert*, discussed *infra*, was rendered under the original 1975 version of the rules. Thus the US Supreme Court's interpretation of Chapter 7 (Opinions Expert and Testimony) was done within the framework of the original promulgation of 1975.

[22] See the discussion in *Experts, Judges, and Commentators: The Underlying Debate About an Expert's Underlying Data*, 47 *Mercer Law Review* 481 (1996); see also the discussion of the US 6th Amendment constitutional right to confrontation and how that impacts on the use of hearsay as a basis for an expert's opinion in *Testimonial Hearsay as the Basis for Expert Opinion: The Intersection of the Confrontation Clause and Federal Rule of Evidence 703 After Crawford v. Washington*, 55 Hastings Law Journal 1939 (2004); and on that issue see also *Expert Evidence and The Confrontation Clause After Crawford. Washington*, 15 *Journal of Law and Policy* 791 (2007).

element of the crime charged or of a defense thereto. Such ultimate issues are matters for the trier of fact alone.

Rule 705. Disclosure of Facts or Data Underlying Expert Opinion. The expert may testify in terms of opinion or inference and give reasons therefor without first testifying to the underlying facts or data, unless the court requires otherwise. The expert may in any event be required to disclose the underlying facts or data on cross-examination.

Rule 706. Court Appointed Experts.

(a) Appointment. The court may on its own motion or on the motion of any party enter an order to show cause why expert witnesses should not be appointed, and may request the parties to submit nominations. The court may appoint any expert witnesses agreed upon by the parties, and may appoint expert witnesses of its own selection. An expert witness shall not be appointed by the court unless the witness consents to act. A witness so appointed shall be informed of the witness's duties by the court in writing, a copy of which shall be filed with the clerk, or at a conference in which the parties shall have opportunity to participate. A witness so appointed shall advise the parties of the witness's findings, if any; the witness's deposition may be taken by any party; and the witness may be called to testify by the court or any party. The witness shall be subject to cross-examination by each party, including a party calling the witness.

(b) Compensation. Expert witnesses so appointed are entitled to reasonable compensation in whatever sum the court may allow. The compensation thus fixed is payable from funds which may be provided by law in criminal cases and civil actions and proceedings involving just compensation under the fifth amendment. In other civil actions and proceedings the compensation shall be paid by the parties in such proportion and at such time as the court directs, and thereafter charged in like manner as other costs.

(c) Disclosure of appointment. In the exercise of its discretion, the court may authorize disclosure to the jury of the fact that the court appointed the expert witness.

(d) Parties' experts of own selection. Nothing in this rule limits the parties in calling expert witnesses of their own selection.

Under the new rules the decision to admit such evidence is again within the discretion of the trial court.[23]

---

[23] Rule 104 FRE states in pertinent part: Preliminary Questions. (a) Questions of admissibility generally. Preliminary questions concerning the qualification of a person to be a witness, the existence of a privilege, or the admissibility of evidence shall be determined by the court, subject to the provisions of subdivision (b). In making its determination it is not bound by the rules of evidence except those with respect to privileges. And see *Daubert, infra,* p. 592 and *General Electric v. Joiner,* 522 U.S. 136; 118 S. Ct. 512; 139 L. Ed. 2d 508 (1997), p. 141 (1997).

After the promulgation of the new rules in 1975, the various states in the United States conducted extensive reviews of their rules and case law. This resulted in almost all of the states adopting the comprehensive language of the 1975 rules as their own rules of evidence. These states, however, continued to apply the *Frye* standard to the admissibility of scientifically based evidence.[24]

The decision in *Daubert v. Merrill Dow*[25]

In June of 1993, the US Supreme Court decided the case of *Daubert v. Merrill Dow*. In that decision, the court, in interpreting the new rules, concluded that the decision in *Frye* no longer exclusively controlled the admissibility of expert testimony and opinion. Instead the court suggested several factors[26] that trial courts should consider when determining the admissibility of scientific evidence in order to test its reliability and accuracy:

Whether the scientific technique or methodology relied upon by the expert to reach his or her opinion:

(1) has been tested or is subject to testing to determine the accuracy of the results. In essence, this factor incorporates the classic "Scientific Method" into the analysis;

(2) has been subject to peer review and publication;

(3) has been examined to determine the technique's known or potential error rate, if any;

(4) has been produced in accordance with existing standards that control the technique's operation;

(5) has been generally accepted in the relevant scientific community as producing accurate results. Note: This is, of course, the admissibility test announced in *Frye*. Thus the result in *Frye* has *not* been completely overruled or abandoned by the Supreme Court; rather, it is now but one of a series of factors which trial courts can consider.[27]

Two subsequent Supreme Court decisions have shed additional interpretive light on the use of the *Daubert* factors, *General Electric v. Joiner*[28] and *Kumho Tire Co. v. Carmichael*;[29] they are briefly discussed below.

---

[24] See, for example, Joseph, G. and Saltzburg, S., The Federal Rules of Evidence in the states, Rule 702, p. 17 (1992) and the discussion in Giannelli, *supra*, §1.11, pp. 68–69.

[25] 509 U.S. 579; 113 S. Ct. 2786; 125 L. Ed. 2d 469 (1993).

[26] But the application of these factors in *Daubert* is flexible and are suggested where appropriate to the facts and are not required in every case, *Kumho Tire v. Carmichael*, 526 U.S. 137; 119 S. Ct. 1167; 143 L. Ed. 2d 238 (1999), pp. 150–151.

[27] See the discussion of the new FRE rules and the decision in *Daubert*, Faigman *et al.*, *supra*, §§ 1-3.0 to 1-3.1, pp. 10–65.

[28] 522 U.S. 136; 118 S. Ct. 512; 139 L. Ed. 2d 508 (1997).

[29] *Kumho, supra.*

In addition, on remand in *Daubert*, the US Ninth Circuit Court of Appeals suggested some other factors which might be of assistance in determining the accuracy of a scientific technique:

(1) whether experts are "proposing to testify about matters growing naturally and directly out of research they have conducted independent of the litigation, or whether they have developed their opinions expressly for purposes of testifying;[30]

(2) whether the expert has unjustifiably extrapolated from an accepted premise to an unfounded conclusion.[31] As noted in *Joiner*,[32] a trial court "may conclude that there is simply too great an analytical gap between the data and the opinion proffered."

(3) whether the expert has adequately accounted for obvious alternative explanations;[33]

(4) whether the expert "is being as careful as he would be in his regular professional work outside his paid litigation consulting".[34]*Daubert* requires that the trial court assure itself that the expert "employs in the courtroom the same level of intellectual rigor that characterizes the practice of an expert in the relevant field";

(5) whether the field of expertise claimed by the expert is known to reach reliable results for the type of opinion the expert would give.[35] *Daubert's* general acceptance factor does not "help show that an expert's testimony is reliable where the discipline itself lacks reliability, as, for example, theories grounded in any so-called generally accepted principles of astrology or necromancy.[36,37]

The scientific method

The scientific method appears to have its foundations in the experiments conducted by Sir Isaac Newton.[38] This protocol is the controlling force[39] behind the determination of the accuracy of any scientific experiment, technique, or process. The scientific method

---

[30] *Daubert*, on remand to the Ninth Circuit US Court of Appeals, 43 F.3d 1311 (1995), p. 1317.

[31] *Daubert*, on remand, 43 F.3d footnote 17, p. 1321.

[32] *Supra.*

[33] Note: The possibility of some uneliminated causes usually presents a question of weight rather than admissibility, so long as the most obvious causes have been considered and reasonably ruled out by the expert.

[34] *Kumho, supra,* p. 152.

[35] *Kumho, supra,* p. 149.

[36] *Kumho, supra,* P. 151; and see *Oglesby v. GMC*, 190 F.3d 244 (4th Cir 1999).

[37] See *Expert Admissibility Symposium: Reliability Standards – Too High, Too Law, or Just Right?*, 33 *Seton Hall Law Review* 987 (2003), for a discussion of some of the effects *of Daubert* in the establishment of a reliability standard.

[38] See also Sir Francis Bacon, *Novum Organum* (1620), which, while not specifically mentioning the "scientific method" as such, focused on scientific research methodology. However, Newton took the issue much further in his *Philosophiae Naturalis Principia Mathematica* (*Mathematical Principles of Natural Philosophy*) (1687).

[39] See, for example, *Daubert, supra,* p. 593, where the Supreme Court stated: "Ordinarily, a key question to be answered in determining whether a theory or technique is scientific knowledge that will assist the trier of fact will be whether it can be (and has been) tested. 'Scientific methodology today is based on generating hypotheses and testing them to see if they can be falsified; indeed, this methodology is what distinguishes science from other fields of human inquiry.'"

follows a series of steps: (1) identify a problem that needs to be solved, (2) formulate a hypothesis, (3) test the hypothesis, (4) collect and analyse the data, and (5) make conclusions.[40] In fact the force of the scientific method in current litigation is such that the Ninth Circuit Court of Appeals, when considering Daubert and applying the Supreme Court's "factors" to the facts in the case on remand, suggested that where there is no scientific consensus among respected, well-credentialed scientists as to what is and is not "good science," the court's responsibility might be to occasionally reject such expert testimony because it was not "derived by the scientific method".[41] The current literature reflects this emphasis as well.[42]

The "pure opinion" rule

There is one other evidentiary concept that should be identified in any discussion of the law of expert testimony in general, the impact of so-called "pure opinion".[43] There exists an important and fundamental conceptual difference between "pure opinion" and the evidentiary requirements of the rulings in either Frye or Daubert that needs to be recognized. Having overcome any preliminary Frye or Daubert challenges,[44] should there be any, the actual opinion of an expert, whether it be based on a set of facts known or made available to the expert or on the results of scientific testing, technology, methodology, or technique, is itself not subject to an analysis under either Frye or Daubert.

> "Pure opinion testimony, such as an expert's opinion that a defendant is incompetent, does not have to meet Frye, because this type of testimony is based on the expert's personal experience and training. While cloaked with the credibility of the expert, this testimony is analyzed by the jury as it analyzes any other personal opinion or factual testimony by a witness."[45]

The expert's opinion under these circumstances rests solely on the credibility of the expert themselves. Thus, aside from any Frye or Daubert challenges as to the accuracy

---

[40] Raum, *supra*, pp. 179–181; Giannelli, *supra*, §1.12, pp. 69–70; and see Faigman *et al.*, §§ 4-1.1.0 to 4-7.0, pp. 115–149, which includes an excellent appendix of further reading on the subject; and see Gauch, H.G., *Scientific Method in Practice*, Cambridge University Press, Cambridge, United Kingdom (2003) p. 456.

[41] *Daubert*, on remand to the Ninth Circuit US Court of Appeals, 43 F.3d 1311 (1995) at p. 1316; Raum, *supra*, pp. 180–181.

[42] For example, see an excellent discussion of the scientific method found in Moenssens, A.A., Henderson, C.E., and Portwood, C.G., *Scientific Evidence in Civil and Criminal Cases*, 5th edn, Chapter 20, sections 20.06–20.09, as it relates to the field of behavioral sciences; and see, for example, the authoritative *Guide for Fire and Explosion Investigations* series by the National Fire Protection Association (NFPA), which commences its in-depth discussion of the topic with an entire chapter offering detailed instruction to investigators on the applicability and use of the scientific method, see Chapter 4, sections 4.1–4.5. Compliance with the procedures in the NFPA 921 guide has formed the basis for the admissibility of scientific fire and arson evidence in cases in the United States.

[43] As it is identified by the courts of the State of Florida.

[44] It is to be noted that this analysis is a two-part process. If any Frye or Daubert challenges have been satisfied or if the underlying scientific technique, etc. has already been deemed to be reliable (i.e., the court takes judicial notice of this fact) then the testimony of the expert is received subject to the limitations described herein. The opinions seem to combine the two-steps, see, for example, *Andries v. Royal Caribbean Cruises Ltd*, 12 So. 3d 260 (3d DCA 2009) and the excellent discussion of the issue in *Marsh v. Valyou*, 977 So. 2d 543 (Fla. 2007).

[45] *Flanagan v. State*, 625 So. 2d 827, 829 n.2 (Fla. 1993), p. 828.

or validity of the underlying science,[46] the court's ability to regulate the receipt of the expert's opinion itself is focused on an examination of the qualifications of the expert as to the subject matter,[47] the need for such testimony,[48] the relevancy of the opinion to the facts in the case,[49] and the relative weight of the probative effect of the testimony versus any potential undue prejudice.[50]

The United Kingdom experience

There is very little "case law" in the United Kingdom regarding the admissibility of expert testimony (which is known in the United Kingdom as "expert evidence"), but what does exist is instructive on the subject.[51] In 1984 the South Australia Supreme Court decided a case[52] that, while it does not adopt *Frye* outright, nonetheless seems to be modeled on the standard announced in *Frye*. The case is now quoted and cited as authority on the subject.[53] The *Bonython* and *Barings* decisions both use a two-part test to determine the admissibility of expert evidence. Under these decisions, the court must:

(1) make a preliminary determination that the proposed expert evidence falls within the customary boundaries of expert evidence. To reach this conclusion the court must (1a) decide whether "the subject matter of the opinion is such that a person without instruction or experience in the area of knowledge or human experience would be able to from a sound judgment on the matter without the assistance of witnesses possessing special knowledge or experience in the area" (the so-called "necessity" or "helpfulness" requirement[54]).[55] This also falls under the category of relevancy.[56] If the answer to this question is in the affirmative, then (1b) the court must determine "whether the subject matter of the [proposed] opinion forms part of a body of knowledge or experience which is sufficiently organised or recognised to be

---

[46] See, for example, FRE 702 "…, (2) the testimony is the product of reliable principles and methods…".

[47] See, for example, FRE 702 "… a witness qualified as an expert by knowledge, skill, experience, training, or education, may testify thereto in the form of an opinion or otherwise (1) If…".

[48] See, for example, FRE 702 "If scientific, technical, or other specialized knowledge will assist the trier of fact to understand the evidence or to determine a fact in issue… ", or the application of *Frye*.

[49] See, for example, *State v. Griswold, supra,* and the excellent discussion of the issue in *Marsh v. Valyou,* 977 So. 2d 543 (Fla. 2007), and FRE 702 "…(3) the witness has applied the principles and methods reliably to the facts of the case."

[50] See FRE Rule 403, which provides: "Although relevant, evidence may be excluded if its probative value is substantially outweighed by the danger of unfair prejudice, confusion of the issues, or misleading the jury, or by considerations of undue delay, waste of time, or needless presentation of cumulative evidence."

[51] See *Folks v Chadd,* discussed *supra.*

[52] *R. V. Bonython* (1984 S.A.S.R.) 45.

[53] See, for example, *Barings Plc v. Coopers & Lybrand,* Chancery Division (Transcript) 9 February 2001, *The Times,* 7 March 2001.

[54] One text refers to this requirement as the "one and only authentic criterion of admissibility", see Roberts, P. and Zuckerman, A., *Criminal Evidence,* 2nd edn (2010), §11.4, p. 486 and the discussion therein.

[55] See, for example, *R. V. Turner* (1975) 1 QB 884. CA and *R. v. Gilfoyle (No 2)* [2001] 2 Cr App Rep 57; *R v Weightman* (1990) 92 Cr App Rep 291, [1991] Crim LR 204; *R. v. Turner (Terence)* (1975) 60 Cr App R 80 [1975] QB 834.

[56] Compare FRE 401 where it states: "'Relevant evidence' means evidence having any tendency to make the existence of any fact that is of consequence to the determination of the action more probable or less probable than it would be without the evidence."

accepted as a reliable body of knowledge or experience,[57] a special acquaintance with which of the witness would render his opinion of assistance to the court."[58] This appears to be simply a rephrasing of the *Frye* standard.[59] If the court finds that the "customary boundaries" test has been met, then

(2) the court should next examine the proposed witness's qualifications to offer the opinion by determining if the witness "has acquired by study or experience sufficient knowledge of the subject to render his opinion of value in resolving the issues before the court".[60]

Having successfully reached this point in the admissibility inquiry, the court must then determine whether the relevancy of the proposed evidence is outweighed by any prejudicial effect that it might have on a jury.[61] This common law concept is embodied in the FRE in Rule 403.[62] If there is no undue prejudice, the last hurdle has been cleared and the expert witness should be allowed to offer their opinion to the jury.

As an aside, the cross-examination of opposing witnesses has had a major role in the trial of a case before the courts of England and the rest of the United Kingdom. Typically cross-examination can be used to successfully expose a witness's bias and to test their ability to observe, hear, and understand external stimuli. With respect to an expert witness, cross-examination can also test the depth and soundness of the witness's qualifications as an expert. However, the inquiry should not stop there. The rules of evidence that we have been discussing have at their roots the imperative that any scientific evidence should first and foremost be accurate and valid. Both the *Frye* and *Daubert* evidentiary thresholds are designed to guarantee, so far as is possible, that the scientific testing, methodologies, and techniques that form the basis for expert opinion produce accurate results if performed correctly and are valid for the purposes to which it was applied. Essentially, *Frye* and *Daubert* permit opposing counsel the opportunity to show a judge and a jury at least a glimpse, and perhaps a long look, into the accuracy and validity of the basis on which the expert witness has formed their opinion. Armed with this information, the jury is arguably in a much better position to gauge the credibility of the expert witness by actually examining the entire process which culminated in the opinion. To this end, despite the fact that neither *Frye* nor *Daubert* nor the US Federal Rules of

---

[57] Roberts and Zuckerman, *supra*, suggest (p. 494) that there is no formal rule as to "field of expertise", see the discussion therein; see *R. v. Dallagher* [2003] 1 Cr App Rep 195 and *Strudwick and Merry* (1993) 99 Cr App R 326, where it was said that "evidence based on a developing new brand of science or medicine is not admissible until accepted by the scientific community as being able to provide accurate and reliable opinion."

[58] *Bonython, supra*, p. 46.

[59] See *R v. Dallagher, supra*, where it stated that the analogy with R 02 is clear; and see *R. v. Gilfoyle* (No 2), *supra*.

[60] For further study the reader should refer to the discussions of this process in *Hodgkinson, supra*, pp. 13–19, and Roberts and Zuckerman, *supra*; and see *R v. Silverlock* (1894) 2 QB 766, *R. v. Robb* (1991) 93 Cr App Rep 161, and *R. V. Dallagher, supra*, in general.

[61] As to evidence for the prosecution, see *R. v. Sang* (1980) AC 402.

[62] FRE 403 provides that: "Although relevant, evidence may be excluded if its probative value is substantially outweighed by the danger of unfair prejudice, confusion of the issues, or misleading the jury, or by considerations of undue delay, waste of time, or needless presentation of cumulative evidence."

Procedure are binding authority in the United Kingdom, there is nothing which prevents counsel from using the *Daubert* standards on cross-examination, where appropriate, to impeach the credibility of an expert witness.[63] In this regard, *Daubert* is not being used on the issue of admissibility of the expert's opinion, but rather to attack the credibility of the expert's opinion and the weight that a jury might give to that opinion.

Additionally, in this area there a few *nisi prius* decisions where trial courts have permitted the use of the *Daubert* factors on cross-examination of an expert witness during the hearing on admissibility of the opinion as well. This was done in an effort to demonstrate that the expert's assertion that the science involved was "generally accepted" could not be valid if the methodologies, etc. could not withstand a *Daubert* attack.[64] In other words, on what was the conclusion of "general acceptance" based?

The criminal procedure rules 2010, s.33

Initially, whenever the evidence of an expert is anticipated, the proponent of that evidence should consult the provisions of The Criminal Procedure Rules 2010 (CrPR) s. 33 for guidance. While all of these rules must be applied, where applicable, to any expert, two provisions of the rules are important for our discussion.

33.2 expert's duty to the court

(1) An expert must help the court to achieve the overriding objective by giving objective, unbiased opinion on matters within his expertise.

(2) This duty overrides any obligation to the person from whom he receives instructions or by whom he is paid.

(3) This duty includes an obligation to inform all parties and the court if the expert's opinion changes from that contained in a report served as evidence or given in a statement.

The requirements of 33.2 are significant in that if an expert departs from this high standard his or her credibility will most certainly be forever tainted.

33.3 content of expert's report

(1) An expert's report must:

   (a) give details of the expert's qualifications, relevant experience and accreditation;

   (b) give details of any literature or other information which the expert has relied on in making the report;

---

[63] For example, see the comment in footnote 11, *Daubert* on remand, 43 F.3d, p. 1319.

[64] See, for example, *Ficic v State Farm Fire & Cas*. Co *et al.*, 9 Misc. 3d 793; 804 N.Y.S.2d 541 (2005) and *Clemente et al. v. Blumenberg*, 183 Misc. 2d 923; 705 N.Y.S.2d 792 (1999). Note: These two decisions were rendered in New York State, which is a jurisdiction controlled by *Frye*.

(c) contain a statement setting out the substance of all facts given to the expert which are material to the opinions expressed in the report, or upon which those opinions are based;

(d) make clear which of the facts stated in the report are within the expert's own knowledge;

(e) say who carried out any examination, measurement, test or experiment which the expert has used for the report and:

   (i) give the qualifications, relevant experience and accreditation of that person,

   (ii) say whether or not the examination, measurement, test or experiment was carried out under the expert's supervision, and

   (iii) summarise the findings on which the expert relies;

(f) where there is a range of opinion on the matters dealt with in the report:

   (i) summarise the range of opinion, and

   (ii) give reasons for his own opinion;

(g) if the expert is not able to give his opinion without qualification, state the qualification;

(h) contain a summary of the conclusions reached;

(i) contain a statement that the expert understands his duty to the court, and has complied and will continue to comply with that duty; and

(j) contain the same declaration of truth as a witness statement.

(2) Only sub-paragraphs (i) and (j) of rule 33.3(1) apply to a summary by an expert of his conclusions served in advance of that expert's report. [Note. Part 27 contains rules about witness statements. Declarations of truth in witness statements are required by section 9 of the Criminal Justice Act 1967 and section 5B of the Magistrates' Courts Act 1980. A party who accepts another party's expert's conclusions may admit them as facts under section 10 of the Criminal Justice Act 1967. Evidence of examinations etc on which an expert relies may be admissible under section 127 of the Criminal Justice Act 2003.]

Effective from 10 April 2010, this new rule governs the required contents of an expert witness's report.[65] A careful reading of this rule reveals virtually everything that a

[65] This rule is in complete conformance with the expert report requirements of ISO/IEC 17025, which is an existing standard governing the establishment and maintenance of a forensic laboratory and is currently in use in the United States and Europe. Compliance with ISO/IEC 17025 is used as a standard in the review process for the certification of forensic laboratories in the United States by ASCLD, The American Society of Crime Lab Directors; see www.ascld.org/ and ISO, International Organization for Standardization, at www.iso.org/iso/catalogue_detail.htm?csnumber=39883 for more details. NOTE: a revision of ASTM standard E620, governing the expert's report was approved in May of 2011 bringing it into agreement with ISO/IEO 17025; see www.astm.org/Standards/E620.htm.

lawyer needs to know preliminarily about an expert's potential evidence. Both the proponent of the expert's evidence as well as any opponents should use this report to begin their education regarding the science involved in the expert's presentation. In this regard, it is critical that the lawyer/proponent of the testimony fully understands the scientific details of the basis for the expert's opinion. This is required not only to properly present the testimony at trial but also to prepare for discovery and depositions, and to rehabilitate the expert on redirect examination as to cross-examination challenges.[66] In the United States, civil cases are very often decided pre-trial by summary judgment based upon an expert witness's deposition. This applies with equal force to the lawyer/opponent so that they may properly challenge any proposed evidence. During the lawyer/expert consultation, the lawyer/proponent should provide the expert with a detailed analysis of the underlying rules of evidence that will control the admissibility of the expert's evidence because it will be up to the expert to convince the trial judge to admit the evidence prior to the witness's actual testimony at trial.[67]

The law commission consultation paper no. 190

As the time of writing, the Law Commission Consultation Paper No 190, The Admissibility of Expert Evidence in Criminal Proceedings in England and Wales, is before the bench and bar for their consideration. This extensive report recommends the codification of the rules regarding the admissibility and use of expert evidence. The report itself appears to be largely based on the successful experience in the United States with its codification of the federal rules of court in 1975, the amendments of 2000, and the adoption of versions of these rules by the various states, with particular reference to Chapter 7 relating to opinion and expert testimony.

---

[66] In the United States, civil cases are very often decided pre-trial by summary judgment based upon an expert witness's deposition.

[67] For an excellent discussion of the lawyer-expert communication issues see Expert Witnessing: Explaining and Understanding Science, Meyer, C., Ed, CRC Press (1999), Chapters 8 & 9.

5 Use and guidelines for plant DNA analyses in forensics

Matthew A. Gitzendanner, Ph.D.

Introduction

Genetic analyses, while perhaps not as common or conclusive as TV shows would suggest (e.g. Baskin and Sommers 2010), have become powerful tools in forensic investigations. DNA evidence has grown tremendously in its use, accuracy, and acceptance since its introduction in the mid-1980s (Butler 2009). This growth, however, has largely been limited to applications of genetic testing of human forensic evidence. While some attention has been paid to non-human samples (e.g. Coyle 2008), genetic testing of plant samples in forensic investigations remains rare. This chapter outlines some of the applications of genetic analysis of botanical forensic samples. In many cases the morphological and anatomical characteristics used to identify a plant sample that are discussed in other chapters will be absent or inconclusive so that a plant sample cannot be identified. In a few cases it may be desirable to identify a particular plant, providing a genetic match between sample and "suspect" or focal plant. Genetic tools can provide data in these cases, greatly increasing the evidentiary value of botanical samples.

Deoxyribonucleic acid (DNA) is the genetic material of an organism. Universal to all life (although some viruses, such as HIV, use RNA instead), DNA is amazing in its simplicity: four bases that, when strung together in a certain order, spell out the blueprint of life. Through its universality, DNA ties all living things together, but also records the history of mutations that change the order of the four bases over time. Studying these genetic differences between individuals is the basis of DNA forensic investigations. Some portions of the genome (all of the DNA in an organism) change more quickly than others, allowing scientists to use different segments of the genome to make comparisons at different levels: across all life, kingdoms, families, species, populations, and related

Forensic Botany: A Practical Guide, First Edition. David W. Hall and Jason H. Byrd.
© 2012 John Wiley & Sons, Ltd. Published 2012 by John Wiley & Sons, Ltd.

individuals. The simplicity of DNA belies the incredible complexity of the stories that it can tell.

Genetic analyses can aid investigators by providing species identification, indicating the population of origin of a sample, and providing evidence of genetic identity between two samples. Depending on the type of sample, the questions needing to be addressed, and the resources available, investigators can select the appropriate genetic analyses for their case. This chapter outlines the main tools available, providing guidelines for collecting and storing samples, as well as a general overview of the various methods that are used frequently. The fields of narcotic plant testing (Coyle *et al.* 2001, 2003), patented cultivar, and genetically modified plant tracking (Gressel and Ehrlich 2002) are intentionally not addressed. The methods provided are not step-by-step laboratory protocols, but are general summaries to assist investigators in understanding the laboratory and analytical methods that could be employed by contracted laboratoriess. By providing an overview of the methods and typical applications of genetic analyses in forensic botany the intention is to provide the non-technical user with the background to understand how properly to collect and store samples to maximize their utility and to have a starting point as they seek to have the samples analysed. A summary of cases in which genetic analyses of plants have been used is provided at the end.

Types of samples and collection for DNA analyses

A wide variety of plant tissues can be used for genetic analysis. Forensic investigations should collect all samples that may be of use and determine their suitability later. A sample not collected because its relevance is overlooked is lost forever. Very little in the way of special tools or storage is needed, and most samples can be inexpensively preserved for long-term storage and later analysis should they become relevant to the investigation. However, the DNA in an improperly preserved specimen can quickly become degraded and its forensic value greatly limited. Proper collection and subsequent storage of samples will increase their utility and prolong the window for successful genetic analysis.

Leaves are generally the preferred tissue for genetic analysis and are what most botanists use in routine studies. If collecting from live plants, one or two healthy, green leaves (depending on size, giving a total of about 2 cm^2) should be chosen. Avoid leaves with evidence of rot, infection, or insects. If the evidence is not from a living plant, or leaves are not available, other tissues can be of use. Dried or dead leaves can be used, however DNA will degrade over time and the success rates of these samples may be quite low (e.g. Craft *et al.* 2007). Flowers, either living or dried, are also suitable samples and can provide good quality DNA for analyses. While decaying wood is unlikely to yield suitable DNA, there have been several recent advances in the use of wood for genetic analysis (Finkeldey *et al.* 2009). DNA yields from wood samples tend to be lower than from leaf tissue and compounds that inhibit polymerase chain reaction (PCR) are often a problem, but methods are improving. Seeds (Walters *et al.* 2006) and fruits, and even roots (Kumar *et al.* 2003) can provide suitable DNA for genetic analyses. In short, almost any plant tissue has potential, although the age, condition, and original DNA content of the tissue will determine its utility. When in doubt, collect the sample.

One of the simplest methods of preserving plant samples for genetic analysis is drying the sample in silica gel (e.g. Fisher Scientific catalog #S684-212). Removing moisture from the sample inhibits the enzymes that would otherwise start to break down the DNA and other cellular components. Dried samples are also less susceptible to fungal growth. Leaves can be placed in a sealed plastic bag (such as a small zipper bag) and a few grams of silica gel added directly to the bag. Mixing some indicating silica gel (e.g. Fisher Scientific catalog #S162-212) will help determine if the silica gel is exhausted and needs replacing. Once the leaves are dry, the sample is stable at room temperature almost indefinitely.

Silica gel storage is applicable to most samples. While a dried leaf or wood sample may be able to survive prolonged storage without silica gel, the addition of silica gel may prolong the length of time DNA can be successfully extracted and the sample will not deteriorate. Samples with exceptionally high moisture content, such as cacti or succulents, may need to have their silica gel replaced several times to achieve sufficient drying.

Room temperature storage without proper desiccation should be avoided if at all possible. DNA is, however, a remarkably resilient molecule and it should never be assumed that a sample will not yield DNA – DNA has been successfully extracted from 150-year-old dried specimens (Rogers and Bendich 1985; Savolainen *et al.* 1995), seeds from archeological sites (Cappellini *et al.* 2010), and even a 17–20-million-year-old fossil (Soltis *et al.* 1992). Experimentation with similarly preserved non-evidence samples becomes more important with marginal samples, but if a reasonable genetic test can assist an investigation, sample quality alone should not discourage attempts to conduct a genetic analysis.

If silica gel desiccation is not practical, samples that will be analysed within a few days can be stored at 4°C (refrigerated). Freezing at $-20°C$ is suitable for samples that need to be stored for a few months, and $-80°C$, ultracold storage will preserve samples for several years. In all cases, samples should be stored in airtight plastic bags or containers. While these methods are feasible for small quantities of samples, freezer space is typically far more limited and costly than room temperature storage, thus desiccation as described above is generally preferred.

Uses of genetic data

Plant species identification

When a sample cannot be positively identified based on morphological or anatomical analyses, because diagnostic characteristics are absent or only small fragments are found, it may be possible to identify a species using DNA sequences. In DNA sequencing the order of the DNA bases of a particular segment of the organism's genome is determined. DNA sequencing of the same segment from different samples allows samples to be compared for similarities and differences. DNA barcoding is a rapidly evolving field in which a standardized set of DNA regions is sequenced and used to identify (often to species) an unknown sample based on comparison to a database of known samples. While barcoding studies in animals have standardized on the cyto-chrome oxidase I gene (www.barcodeoflife.org), for plants the best regions (genes or portions of the genome) for barcoding are still under debate (e.g. Rubinoff *et al.* 2006; Kress *et al.* 2009; Roy *et al.* 2010). However, there is general consensus that some combination of *rbc*L, *mat*K, the *trn*H-*psb*A spacer, and the internal transcribed spacer (ITS) sequences will generally provide a good identification at the species level (Kress *et al.* 2005; Rubinoff *et al.* 2006; Valentini *et al.* 2009; Dunning and Savolainen 2010; Roy *et al.* 2010). Given that relatively few samples will typically be analysed, the recommendation would be to sequence all four of the above loci as this would increase the chances of clear species identification. The basic process involves extracting the DNA from the sample, PCR amplifying the loci with conserved primers, sequencing these regions with Sanger sequencing, and comparing the sequences to the ever growing collection of sequences in public databases like the National Center for Biotechnology Information GenBank database (see www.barcodeoflife.org for details of the process). DNA barcode identifications can then be used in much the same way as the species identifications described in the rest of this book: providing evidence of habitats, associated plants, etc.

Nearly universal PCR primers for each barcoding locus are available, although it is recommended that investigators search the literature as primer sequences are constantly being refined and many loci require primer pairs that are specific to related plant species or higher taxonomic groups such as families or orders (e.g. *mat*K is best amplified with different primers for different plant orders (Dunning and Savolainen 2010)). The amplification and sequencing methods in Kress *et al.* (2005) should provide a good source for barcoding studies. Current costs for this type of analysis will average from $15 to $40 per sample depending on the number of loci sequenced and the quality of the sample. Substantial additional costs may be incurred if the laboratory facility charges processing fees.

DNA barcoding will greatly expand the utility of fragmentary samples, opening up new possibilities in forensic botany. However, this also means that investigators must be aware of the potential, and collect and preserve these samples during investigations. Multiple fragmentary samples that once may have provided little information for a forensic botanist may now provide multiple species identifications, painting a picture of a habitat and potentially offering a critical clue to investigators.

Identification of population of origin

Identifying the population, or general region, of origin of a plant sample can frequently be accomplished with a combination of the sequence-based methods described above and the marker-based methods described below. Phylogenetic analysis

Plate 1.8 The amount of wilt present on broken or damaged leaves and branches can help to establish a time frame. (Courtesy of Dr James L. Castner.)

Plate 1.9 Chlorophyll degradation of plant leaf from a burial site can be used for time interval estimations. (Courtesy of Dr J. H. Byrd.)

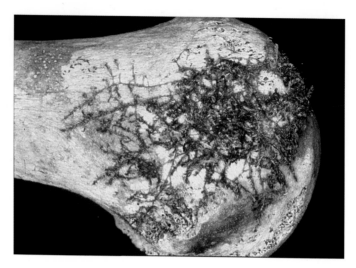

Plate 2.15 A moss growing on human bone can useful for determining a portion of the postmortem interval.

Plate 2.16 The presence of algae can link individuals and items to particular bodies of water. A blue-green alga, commonly known as "pond scum", is shown.

Plate 2.17   A fungus, commonly known as a "rust", on the trunk of a pine tree.

Plate 2.18   Shelf fungus, common in forested environments, can be utilized to provide time intervals based on its rate of growth.

Plate 2.21   A fruticose lichen is branched and easily removed from items.

Plate 2.46   Anthers covered with pollen.

Plate 3.18 Cut roots define the edge of the original burial pit in this excavation of a clandestine gravesite.

Plate 6.4 Various plant species have distinctive hairs, or trichomes, on their leaves, and the stereomicroscope is useful for examining these. Scale bars = 1 cm in A and C, 100 microns in B and D. (*See text for full caption.*)

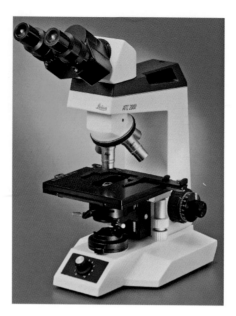

Plate 6.5 The compound microscope is used for examining microscopic detail on fiber or cellular/tissue-level detail of plant structures thin enough to let light pass through them. The Leica ATC-2000 is an education-grade microscope. Photo courtesy of Leica Microsystems, Inc., Bannockburn, Illinois, USA.

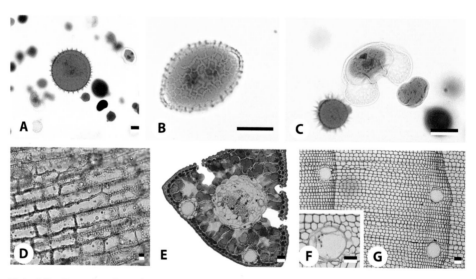

Plate 6.7 Examples of specimens typically examined with the compound microscope. (*See text for full caption.*)

Plate 9.3 Macroalgae are algae with bodies large enough to be seen to the naked eye. (*See text for full caption.*)

Plate 10.5 This skeleton shows significant taphonomic change due to weathering from environmental exposure. (Photograph Courtesy of C. A. Pound Human Identification Laboratory, University of Florida). (*See text for full caption.*)

Plate 10.7 Plants fragments recovered from the hair of a sexual assault victim allowed investigators to pinpoint the site of the attack. (Courtesy of Dr. J. H. Byrd)

Plate 10.11 Chlorosis is a term used to describe the death of cholorophyll and yellowing of green plant leaves. (*See text for full caption.*)

(i.e., reconstructing the evolutionary history of the samples) of DNA sequences from samples from a broad range of a species can frequently identify patterns differentiating regions (for a review see Petit and Vendramin 2007). For finer scale analyses, population-level studies using microsatellites or the other markers outlined below can similarly be used to determine the population of origin. Genetic marker data can be analysed to either probabilistically assign a sample of unknown origin to sampled populations or statistically exclude an unknown sample as having originated from sampled populations (e.g., Manel *et al.* 2002, 2005).

Identification of individual plants

The backbone of human forensic investigation involves providing a genetic match between collected evidence and a sample in a database, which is in turn linked to a person. The details of the genetic markers used will be discussed later, but the success of human forensic studies relies on a large database of genetic data genotyped with the same set of loci from millions of individuals (Butler 2009). In the early stages of human genetic analyses, different jurisdictions used different loci, preventing aggre-gation and comparisons across jurisdictions (different countries still use different systems and have different laws governing how samples can be added and used (Jobling and Gill 2004)). The standardization on the Combined DNA Index System (CODIS) loci and the establishment of the National DNA Index in 1994 were great advances. Using the CODIS markers and the database of genotypes, human forensic scientists can rapidly find exact matches and, just as importantly, assess of the chances of a random match.

Random matches can and do occur at any particular locus. For example, almost one in four individuals has the 15 repeat allele at the D3S1358 human forensic locus (Butler *et al.* 2003). However, when multiple loci are studied, the probabilities are multiplied, so that a one in four chance at one locus and one in 50 at a different locus combine to one in 200 with both loci. Adding the full set of 13 loci commonly used in the human forensic lineup, probabilities of a genotype rapidly fall to less than 1 in 7 billion, the number of people in the world – strong evidence that the match is not just due to chance!

The power of plant forensic investigations is limited by the lack of a CODIS-like database. A match between evidence and focal plant means nothing without informa-tion on how likely a match is due to chance alone – can the markers used distinguish individual plants? Distinguishing individual plants may be more difficult if the plant naturally reproduces clonally, or is clonally propagated and the same genotype planted throughout the country. A forensic investigation that seeks to match evidence to a particular plant will require building a database of samples. This will involve collecting and genotyping many samples from near the crime scene, or focal plant, and other places where the plant occurs. By genotyping samples near and far, investigators can determine the likelihood of chance matches. This is a sizable undertaking and it is far easier to demonstrate that two samples do not match than to convincingly show that they do match and that the match is not due to chance alone.

Before embarking on a genetic investigation that seeks to provide a match between two samples, it is important to consider several factors that can predict potential outcomes. This may guide investigators who may have multiple species to choose from in selecting which ones to pursue genetic tests on. Investigations of common plants occurring across a wide range will be much less likely to be successful than an investigation of a rarer plant found only infrequently. Annual plants will be difficult to match unless the investigation proceeds quickly and samples from one growing season can be collected. Wind-pollinated species tend to be less genetically differentiated than insect- or bird-pollinated species. There are several environmental correlates to genetic differentiation (Hamrick and Godt 1996), and using these to select target species may improve the likelihood of success.

Genotyping methods

General considerations

Regardless of the markers employed, the same basic methods and precautions will be necessary. Once a determination has been made that genetic data will sufficiently strengthen a case and a laboratory has been identified to conduct the analyses, a genetic marker must be selected. This choice will be dictated based on the expertise of the laboratory, the goals of the analysis, and budget constraints. Microsatellites are currently the favored genetic markers for a wide range of analyses; however, as outlined below, the costs can be quite high and the goals may be more easily achieved with other markers.

Once a marker system has been selected, it is important to establish that the laboratory is able to successfully genotype samples similar to those that constitute the case sample(s). Freshly collected green leaves of the same species are the best place to start with optimizing protocols, but it is important at this stage to establish that the protocols work on samples as similar to the case sample(s) as possible. If the evidence is leaf debris, leaf debris of similar age should be analysed. Using non-evidence samples, the researchers can modify and optimize protocols to increase the probability of success once the evidence is analysed. At this stage it is also important to establish that the marker selected has the variation and resolution necessary to address the question at hand. If these initial experiments show little or no genetic variation across a wide range, a different marker may need to be selected.

Once the laboratory has established the protocols for the genotyping of the evidence and has shown that the markers can address the question at hand, the study of the evidence can proceed. Great care must be taken to ensure the cleanliness of the laboratory and prevent cross-contamination of samples. If the goal is to link evidence collected at a crime scene to a particular plant, samples from the plant and the evidence should not be handled at the same time. Ensuring separation of the two reduces the possibility of cross-contamination.

If a sufficient amount of the evidence is available, two to three sub-samples should be taken and analysed independently – this is especially critical for the anonymous markers discussed below. Even when only one DNA extraction can be made from the sample, it should be processed through the protocol multiple times, again providing replicate genotypes and ensuring repeatability of genotypes. All of the methods discussed below rely on the PCR to generate the fragments that will be genotyped. PCR is sensitive to secondary compounds and fragmented DNA, and can result in what is known as allelic dropout, where a fragment does not amplify from a sample, not because the allele is not present, but because of failure of PCR. Multiple reactions and careful validation can compensate for this.

DNA extraction

Most plant DNA extraction methods involve a disruption of the tissue, addition of a buffer containing detergents to break open cells and membranes, removal of unwanted compounds, and a process to precipitate or filter the DNA out of solution for purification. One common method, first outlined by Doyle and Doyle (1987) and modified by Cullings (1992), uses a cetyltrimethylammonium bromide (CTAB) buffer and chloroform extraction followed by isopropanol precipitation. Many prefer commercial kits, most commonly the Qiagen DNeasy Plant mini extraction kits (Qiagen Inc., Valencia, CA). Modifications to these protocols have been published for different tissues (e.g. Cheng *et al.* 1997) or difficult species (e.g. Porebski *et al.* 1997). While the kit-based methods offer more standardization and sometimes yield higher quality DNA, kits are less amenable to modifications for dealing with difficult samples. In the end, the extraction method will depend on the preference of the laboratory contracted to do the work and the success of trial investigations on samples similar to the evidence.

Microsatellites

Human fingerprinting is largely based on a set of 13 microsatellite loci. These are segments in the genome where two or three nucleotides are repeated over and over again in a string (e.g. ATGATGATGATGATGATG). Microsatellite regions are known from all eukaryotes and have been shown to evolve very rapidly by adding or subtracting repeat units (Li *et al.* 2002; Richard *et al.* 2008). This means that individuals within a species, population and even family will frequently have different numbers of repeat units. Each locus is amplified using PCR primers designed specifically for that locus. The product is then sized, and the number of repeats calculated and compared to the other samples or a database of known samples.

One challenge when applying microsatellites in plant forensic investigations is that in most cases there are no microsatellite loci available for the investigation. Loci tend to be species-specific, so each investigation will require an initial investment in developing loci for the species of interest. Luckily, locus development has become easier over time, and commercial facilities can generally be hired to develop loci for $2000–$5000. Additionally, as locus development becomes easier, researchers are publishing more studies with loci that could be used, so a literature search may find suitable loci for an investigation. Once loci are available, genotyping samples involves DNA extraction, PCR amplification, sizing, and analysis. In addition to the development of the loci, a microsatellite study should budget $10–$15 per sample for analysis of the samples.

Random/anonymous markers

While microsatellite loci are generally species-specific and require an initial investment in developing the loci, random or anonymous markers offer methods more easily adapted to new species. However, these can be more problematic. The name of this class of markers comes from their reliance on amplifying random or anonymous portions of the genome. By surveying enough random locations in the genome, individual samples can be distinguished. However, the basis of differences is often not known and several genetic changes can account for allelic differences. Each method has it own benefits and problems.

The most common anonymous markers are randomly amplified polymorphic DNA (RAPD) (Williams *et al.* 1990), inter-simple sequence repeats (ISSRs) (Zietkiewicz *et al.* 1994), and AFLPs (Vos *et al.* 1995). The reliability of these markers has been a major issue (Penner *et al.* 1993; Perez *et al.* 1998; Bagley *et al.* 2001) and many scientific journals will no longer publish scientific studies that use datasets based entirely on RAPD or AFLP data (e.g., *Molecular Ecology* and *Conservation Genetics*). Despite these issues, they can be valuable molecular markers for forensic investigations, but investigators should consider how the data will be used and how a cross-examination of the data would stand up.

RAPD, perhaps the least repeatable of the three methods discussed here, is based on amplification of anonymous pieces of DNA using short (usually 10 base pairs (bp)), random primers (Williams *et al.* 1990). Samples are genotyped by running fragments on agarose gels, sizing the fragments, and scoring the presence or absence of fragments of each size across samples. Mutations that cause a gain or loss of a priming site, as well as insertions or deletions between priming sites, create variation for the presence/absence of a band of a particular length. Like all three anonymous markers described here, RAPD is a dominant marker and heterozygotes cannot be detected – individuals with a diploid genotype of either present/present or present/absent will be scored as present. This limits some of the genetic interpretations and means that more loci must be studied relative to microsatellite loci.

ISSRs (Zietkiewicz *et al.* 1994) take advantage of the high frequency of microsatellite loci in genomes and the method is based on using a primer that consists of a microsatellite repeat. The repeat primer is used to amplify the region between microsatellite loci. This differs from microsatellite analysis, which uses primers in the flanking region and surveys changes in repeat numbers in the microsatelite locus. Here the sources of variation are gains and losses of priming sites and changes in the distance between microsatellite loci. Again the method involves DNA extraction, PCR amplification, sizing the fragments, and scoring the presence or absence of bands across samples.

Lastly, AFLP (Vos *et al.* 1995) is based on amplifying fragments of the genome bounded by restricton enzyme recognition sites. The first step in the AFLP process is to digest the DNA with two enzymes that recognize specific DNA sequences in the genomes. These recognition sites are typically 4–6 bp in length and occur more or less randomly throughout the genome. Once digested, adapters of known sequence can be attached to the ends of the fragments and used as PCR primers. Typically two rounds of selective PCR are conducted, each adding one or more arbitrary bases to the primer to reduce the number of fragments amplified to a manageable number. Unlike RAPD and ISSRs, which can typically be scored on agarose gels, AFLP is generally run on a DNA sequencer, similar to microsatellites. Similar to RAPD, the source of variation is the gain or loss of restriction sites and insertions and deletions between them. AFLP is typically considered to be more repeatable than RAPD, as it uses more specific enzymes and more stringent PCR conditions. Meudt and Clarke (2007) provide a recent review of the current best practices for using AFLP, and data scoring and analysis.

Given the issues with reproducibility, why use these methods? Mainly because they are the easiest, most cost-effective way to learn about genetic variation in a set of samples. Costs are close to $10 per sample and there is little or no investment in development. With careful replication, the reliability issues can be addressed, for example by scoring only fragments that are repeatable across replicates. One additional caution, however, is that the ease of applying these methods to any species has a drawback in that they will amplify loci from multiple species in a mixed sample. A leaf with fungal contamination will amplify loci from both the plant and the fungus, and the researcher cannot distinguish the two sources for a band. Since the comparison is between the presence and absence of a fragment of DNA, the extra loci from contaminating sources will lead to false rejection of a match.

Genetic interpretation

Once the data have been gathered, there are several ways to analyse them. At the most basic level, band-for-band matching between evidence and focal plant will provide a genetic match. However, it is important to establish that a match has some statistical significance and cannot be attributed to chance alone.

Match probabilities

Of fundamental importance in studies seeking to show a genetic match between evidence and focal plant is calculation of the match probability. As alluded to above, this is based on the allele frequencies at each locus and is calculated for each genotype by multiplying the allele frequencies found at each locus. This is, however, an oversimplification and ignores the effects of population structure and linkage disequilibrium (Lewontin and Hartl 1991). While much of the literature addressing this is based on human forensics, for which there is a wealth of data, population geneticists and molecular ecologists have also tackled the issue (Waits *et al.* 2001). There are software packages that can assist researchers in making the corrected calculations (e.g., Valière 2002). In addition, the program GIMLET has algorithms to help detect allele dropout and other genotyping errors (Valière 2002). Forensic investigations should not rely solely on these dropout detection methods however, and for particularly difficult evidence samples allele dropout should be expected and accounted for in searching for matches.

Clustering and population assignment

While the typical approach in DNA forensic investigations is to link evidence to one plant, this may not always be the case. In many situations there may be no single focal plant. In these cases, clustering and population assignment methods may still provide credible and valuable forensic information. Clustering methods tend to be based on pairwise genetic distances (calculated in one of several ways) and cluster samples are based on overall genetic similarity. As with the moss example below, this may allow investigators to say that a sample most likely came from a certain population (Craft *et al.* 2007). Similarly, more recent methods, typically referred to as population assignment methods, allow samples to be statistically assigned to populations independent of prior assumptions of populations structure (Pritchard *et al.* 2000; Gao *et al.* 2007; François and Durand 2010). Samples from various locations can be analysed, along with the evidence sample, to determine the most likely population of origin of the sample.

Finding a laboratory for analysis

Knowing of the possibility of genetic analysis of plant samples is only the first step. It can often be difficult to find a laboratory to conduct the analysis. Human forensic laboratories are generally unwilling and unable to analyse plant samples. There are a few commercial and academic laboratories with experience in handling non-human samples, and some have even worked with plant samples. Another option is to recruit laboratories from local or state universities. Many researchers or core facilities at universities would be willing, if asked, to provide genetic analysis services at a reasonable cost. However, it is important be clear on the expectations for sample handling and the chain of custody procedures since few laboratories of this nature will have had forensic experience.

Case studies

What is believed to have been the first use of plant DNA evidence in a US trial was a 1992 murder case in Arizona in which seed pods of a Palo Verde tree (*Cercidium* sp.) were found in the bed of the suspect's pickup truck (Yoon 1993). RAPD analyses showed these pods to be a genetic match with a tree near the body. The RAPD analyses were able to identify individual plants, correctly identifying the focal plant from a line-up of 11 other trees from near the crime scene and 18 additional trees of the same species from outside the crime scene (Yoon 1993).

RAPD and ISSR data were used in a Finnish study linking three suspects to a murder scene (Korpelainen and Virtanen 2003). The suspects were last seen leaving a café with the victim, who was later found dead. Moss samples were found in the suspects' shoes, clothes, and car. The RAPD and ISSR analyses, while not finding exact matches between the samples and moss patches found near the body, did indicate that the samples likely came from the same populations. The moss samples were identified as three different species, although only two were analysed genetically as the third, *Ceratodon purpureus*, reproduces sexually and this was thought to pose problems for the analysis. Three additional samples of *Brachythecium albicans* and seven of *Calliergonella lindbergii* were collected and analysed from the vicinity of the body, along with 16 and 7 samples, respectively, from southern Finland. RAPD and ISSR analyses detected higher levels of variation than expected and no exact matches were found. However, the samples from the suspects matched more closely than samples from the region.

Craft *et al.* (2007) provide an excellent overview of applying microsatellite data to a forensic case. The case involved a double homicide where a pregnant woman and her unborn fetus were murdered in central Florida. Their bodies were found in a shallow grave with three large sand live oak (*Quercus geminata*) trees overhanging the site. The expert botanist consulted on the case identified two dried leaves found in the suspect's car as *Q. geminata* and a third as another closely related oak. Craft *et al.* (2007) attempted to use microsatellites to compare the leaves from the car with the trees at the burial site. As mentioned above, one of the limitations to using microsatellite studies for plants is the relatively few species for which loci are readily available. Craft *et al.* (2007) tested, and successfully amplified, loci that had been developed for *Q. macrocarpa* and *Q. petraea*, showing that, in some cases, loci can be successfully applied across species. Craft *et al.* (2007) analysed samples from 24 *Q. geminata* trees collected within 10 km of the burial site and were able to genetically distinguish each of these samples using four microsatellite loci. Based on the sample, and the allele frequencies at each locus, the average probability of identity was calculated as 2.06×10^{-6} (Craft *et al.* 2007), meaning that about one in 500,000 trees would share a genotype. While this is not particularly high, Craft *et al.* (2007) did not attempt to add more loci as the results with the analyses of the samples in evidence indicated that they were not a match to the trees at the burial site. While Craft *et al.* (2007) were able to amplify some loci from the leaves from the suspect's car, they were not able to amplify all loci. However, those loci that did amplify indicated that the samples were not a match. This study, while not supporting a conviction, should serve as a model for the process involved in applying genetic analyses to botanical forensics.

Lastly, in 2000 I worked with the Spokane County, WA, Sheriff's office to conduct genetic testing of samples of a honey locust tree (*Gleditsia triacanthos*) collected at the burial site of a homicide victim and compare these samples to a tree in the suspects yard.

AFLPs were selected for the study, as that was the marker that our laboratory was using at the time. The commonly planted thornless honey locust is a horticultural variety that is clonally propagated and all samples from the neighborhood around the suspect's home were genetically identical. This is a case where a genetic match of the samples is meaningless.

Conclusions

Genetic analyses of plant samples remain relatively rare. Advances in laboratory methods and genotyping technologies are, however, making these studies more amenable, and it is likely that there will be a growth in their application in the coming decade. Species identification through DNA barcoding is potentially applicable to a much larger set of samples that otherwise lack diagnostic characters for morphological and anatomical methods, opening a new set of possibilities for forensic investigations.

References

Bagley, M.J., Anderson, S.L., and May, B. (2001) Choice of methodology for assessing genetic impacts of environmental stressors: polymorphism and reproducibility of RAPD and AFLP fingerprints. *Ecotoxicology*, 10(4), 239–244.

Baskin, D. and Sommers, I. (2010) The influence of forensic evidence on the case outcomes of homicide incidents. *Journal of Criminal Justice*, 38(6), 1141–1149.

Butler, J.M. (2009) *Fundamentals of Forensic DNA Typing*. Academic Press, Burlington, MA.

Butler, J.M., Schoske, R., and Vallone, P.M. (2003) Allele frequencies for 15 autosomal STR loci on US Caucasian, African American, and Hispanic populations. *Journal of Forensic Sciences*, 48, 908–911.

Cappellini, E., Gilbert, M.T.P., Geuna, F., Fiorentino, G., Hall, A., Thomas-Oates, J., Ashton, P.D., Ashford, D.A., Arthur, P., Campos, P.F., Kool, J., Willerslev, E., and Collins, M.J. (2010) A multidisciplinary study of archaeological grape seeds. 7(2) 205–217.

Cheng, F.S., Brown, S.K., and Weeden, N.F. (1997) A DNA extraction protocol from various tissues in woody species. *HortScience*, 32(5), 921–922.

Coyle, H.M. (2008) *Nonhuman DNA typing: theory and casework applications*. CRC Press, Boca Raton, FL.

Coyle, H.M., Ladd, C., Palmbach, T., and Lee, H.C. (2001) The green revolution: botanical contributions to forensics and drug enforcement. *Croatian Medical Journal*, 42(3), 340–345.

Coyle, H.M., Palmbach, T., Juliano, N., Ladd, C., and Lee, H.C. (2003) An overview of DNA methods for the identification and individualization of marijuana. *Croatian Medical Journal*, 44(3), 315–321.

Craft, K.J., Owens, J.D., and Ashley, M.V. (2007) Application of plant DNA markers in forensic botany: Genetic comparison of Quercus evidence leaves to crime scene trees using microsatellites. *Forensic Science International*, 165(1), 64–70.

Cullings, K.W. (1992) Design and testing of a plant-specific PCR primer for ecological and evolutionary studies. *Molecular Ecology*, 1(4), 233–240.

Doyle, J.J. and Doyle, J.L. (1987) A rapid DNA isolation procedure for small quantities of fresh leaf tissue. *Phytochemistry Bulletin*, 19, 11–15.

Dunning, L.T. and Savolainen, V. (2010) Broad-scale amplification of matK for DNA barcoding plants, a technical note. *Botanical Journal of the Linnean Society*, 164(1), 1–9.

Finkeldey, R., Leinemann, L., and Gailing, O. (2009) Molecular genetic tools to infer the origin of forest plants and wood. *Applied Microbiology and Biotechnology*, 85(5), 1251–1258.

François, O. and Durand, E. (2010) Spatially explicit Bayesian clustering models in population genetics. *Molecular Ecology Resources*, 10(5), 773–784.

Gao, H., Williamson, S., and Bustamante, C.D. (2007) A Markov chain Monte Carlo approach for joint inference of population structure and inbreeding rates from multilocus genotype data. *Genetics*, 176 (3), 1635–1651.

Gressel, J. and Ehrlich, G. (2002) Universal inheritable barcodes for identifying organisms. *Trends in Plant Science*, 7(12), 542–544.

Hamrick, J.L. and Godt, M.J.W. (1996) Effects of life history traits on genetic diversity in plant species. *Philosophical Transactions of the Royal Society of London Series B: Biological Sciences*, 351, 1291–1298.

Jobling, M.A. and Gill, P. (2004) Encoded evidence: DNA in forensic analysis. *Nature Reviews Genetics*, 5(10), 739–751.

Korpelainen, H. and Virtanen, V. (2003) DNA fingerprinting of mosses. *Journal of Forensic Sciences*, 48(4), 804–807.

Kress, W.J., Wurdack, K.J., Zimmer, E.A., Weigt, L.A., and Janzen, D.H. (2005) Use of DNA barcodes to identify flowering plants. *Proceedings of the National Academy of Sciences of the United States of America*, 102(23), 8369–8374.

Kress, W.J., Erickson, D.L., Jones, F.A., Swenson, N.G., Perez, R., Sanjur, O., and Bermingham, E. (2009) Plant DNA barcodes and a community phylogeny of a tropical forest dynamics plot in Panama. *Proceedings of the National Academy of Sciences*, 106(44), 18621–18626.

Kumar, A., Pushpangadan, P., and Mehrotra, S. (2003) Extraction of high-molecular-weight dna from dry root tissue of *Berberis lycium* suitable for RAPD. *Plant Molecular Biology Reporter*, 21 (September), 309a–309d.

Lewontin, R. and Hartl, D. (1991) Population genetics in forensic DNA typing. *Science*, 254(5039), 1745–1750.

Li, Y., Korol, A.B., Fahima, T., Beils, A., and Nevo, E. (2002) Microsatellites: genomic distribution, putative functions and mutational mechanisms: a review. *Molecular Ecology*, 11 (12), 2453–2465.

Manel, S., Berthier, P., and Luikart, G. (2002) Detecting wildlife poaching: identifying the origin of individuals with Bayesian assignment tests and multilocus genotypes. *Conservation Biology*, 16, 650–659.

Manel, S., Gaggiotti, O.E., and Waples, R.S. (2005) Assignment methods: matching biological questions with appropriate techniques. *Trends in Ecology and Evolution*, 20, 136–142.

Meudt, H.M. and Clarke, A.C. (2007) Almost forgotten or latest practice? AFLP applications, analyses and advances. *Trends in Plant Science*, 12(3), 106–117.

Penner, G.A., Bush, A., Wise, R., Kim, W., Domier, L., Kasha, K., Laroche, A., Scoles, G., Molnar, S.J., and Fedak, G. (1993) Reproducibility of random amplified polymorphic DNA (RAPD) analysis among laboratories. *Genome Research*, 2(4), 341–345.

Perez, T., Albornoz, J., and Dominguez, A. (1998) An evaluation of RAPD fragment reproducibility and nature. *Molecular Ecology*, 7(10), 1347–1357.

Petit, R. J. and Vendramin, G.G. (2007) Plant phylogeography based on organelle genes: an introduction. In: Weiss, S. and Ferrand, N. (eds), *Phylogeography of southern European refugia: evolutionary perspectives on the origins and conservation of European biodiversity*. Springer, Berlin, 23–97.

Porebski, S., Bailey, L.G. and Baum, B.R. (1997) Modification of a CTAB DNA extraction protocol for plants containing high polysaccharide and polyphenol components. *Plant Molecular Biology Reporter*, 15(1), 8–15.

Pritchard, J.K., Stephens, M. and Donnelly, P. (2000) Inference of Population Structure Using Multilocus Genotype Data. *Genetics*, 155(2), 945–959.

Richard, G.-F., Kerrest, A., and Dujon, B. (2008) Comparative genomics and molecular dynamics of DNA repeats in eukaryotes. *Microbiology and Molecular Biology Reviews*, 72(4), 686–727.

Rogers, S.O. and Bendich, A.J. (1985) Extraction of DNA from milligram amounts of fresh, herbarium and mummified plant tissues. *Plant Molecular Biology*, 5(2), 69–76.

Roy, S., Tyagi, A., Shukla, V., Kumar, A., Singh, U.M., Chaudhary, L.B., Datt, B., Bag, S.K., Singh, P.K., Nair, N.K., Husain, T., Tuli, R. (2010) Universal plant DNA barcode loci may not work in complex groups: a case study with Indian berberis species. In: Joly, S. (ed.) *PLoS ONE*, 5 (10), e13674.

Rubinoff, D., Cameron, S., and Will, K. (2006) Are plant DNA barcodes a search for the Holy Grail? *Trends in Ecology and Evolution*, 21(1), 1–2.

Savolainen, V., Cuénoud, P., Spichiger, R., Martinez, M.D.P., Crèvecoeur, M., and Manen, J.-F. (1995) The use of herbarium specimens in DNA phylogenetics: Evaluation and improvement. *Plant Systematics and Evolution*, 197(1–4), 87–98.

Soltis, P.S., Soltis, D.E., and Smiley, C.J. (1992) An rbcL sequence from a Miocene Taxodium (bald cypress). *Proceedings of the National Academy of Sciences*, 89(1), 449–451.

Valentini, A., Pompanon, F., and Taberlet, P. (2009) DNA barcoding for ecologists. *Trends in Ecology and Evolution*, 24(2), 110–117.

Valière, N. (2002) GIMLET: a computer program for analysing genetic individual identification data. *Molecular Ecology Notes*, 2(3), 377379.

Vos, P., Hogers, R., Bleeker, M., Reijans, M., van de Lee, T., Hornes, M., Frijters, A., Pot, J., Peleman, J., and Kuiper, M. (1995) AFLP: a new technique for DNA fingerprinting. *Nucleic Acids Research*, 23(21), 44074414.

Waits, L.P., Luikart, G., and Taberlet, P. (2001) Estimating the probability of identity among genotypes in natural populations: cautions and guidelines. *Molecular Ecology*, 10(1), 249–256.

Walters, C., Reiley, A.A., Reeves, P.A., Baszczak, J., and Richards, C.M. (2006) The utility of aged seeds in DNA banks. *Seed Science Research*, 16(03), 169–178.

Williams, J.G.K., Kubelik, A., Livak, K., Rafalski, J., Tingey, S. (1990) DNA polymorphisms amplified by arbitrary primers are useful as genetic markers. *Nucleic Acids Research*, 18(22), 6531–6535.

Yoon, C.K. (1993) Forensic science. Botanical witness for the prosecution. *Science*, 260(5110), 894–895.

Zietkiewicz, E., Rafalski, A., and Labuda, D. (1994) Genome fingerprinting by simple sequence repeat (SSR)-anchored polymerase chain reaction amplification. *Genomics*, 20(2), 176–183.

6 A primer on forensic microscopy

Christopher R. Hardy, Ph.D.

Microscopes and microscopic botanical structures relevant to forensic botany

Microscopes are standard equipment in forensic laboratories since much evidence is either microscopic or fragmentary in nature and requires the observation of microscopic features for positive identification. Of the wide variety of microscopes employed in the plant sciences, however, only a small subset is needed for most forensic applications. In the following paragraphs, a liberal concept of what constitutes a microscope is adopted to include any magnification aid commonly employed in forensic botany.

The hand-lens (also known as a loupe)

The simplest, lowest powered, least expensive, and most frequently employed microscope is the hand-lens (Figure 6.1). Hand-lenses are small, hand-held, monocular magnification tools used to magnify objects generally from 3X to 20X magnification. The power of a hand-lens is fixed with the lens, and the standard magnification used by botanists is 10X, although 15X and 20X lenses are sometimes used. Whereas most have just one magnification available per unit, others have two or three lenses – all with different magnifications – available. Unlike the familiar magnifying glasses used by librarians or other readers of books, hand-lenses do not have a handle *per se*. Jewelers use hand-lenses to examine diamonds and gemstones for various qualities, geologists use them to examine minerals, gems, and rocks, while botanists use them to examine very small parts, hairs (trichomes), or other minute surface features of flowers, leaves, stems, and roots or their parts. Indeed, the hand-lens is not powerful enough to see truly

Forensic Botany: A Practical Guide, First Edition. David W. Hall and Jason H. Byrd.
© 2012 John Wiley & Sons, Ltd. Published 2012 by John Wiley & Sons, Ltd.

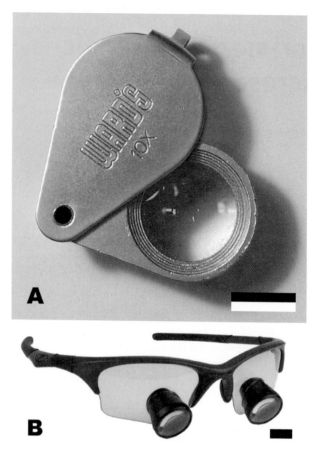

Figure 6.1 The hand-lens, or loupe, similar to the one depicted in A, is the simplest, lowest-powered, least expensive, yet perhaps most broadly useful microscope available to the forensic botanist. Hands-free alternatives such as the one depicted in B offer two lenses attached to eyeglasses. These are more expensive than the hand-lens, but may be favored by coroners or others requiring low-magnification while performing precision tasks. Scale bars = 1 cm. (Photograph of the SurgiTel® Micro 250 in B courtesy of Daniel R. Hardy & SurgiTel®, Ann Arbor, Michigan, USA.)

microscopic structures, but the forensic investigator can use it to examine potential physical evidence (e.g., an article of clothing) to determine whether or not plant matter is present, and, with the proper experience, to identify the plant species to which a piece of trace evidence belongs.

The utility of hand-lenses over other magnification tools lies in their low cost (generally US$5–30) and portability (small and lightweight), which means that they can be used directly in the field or extemporaneously at some location where there is not a microscope available. Moreover, they are easily purchased in any university bookstore, gems store, some jewelry supply stores, museum souvenir shops, or from common general merchandise sites on the internet.

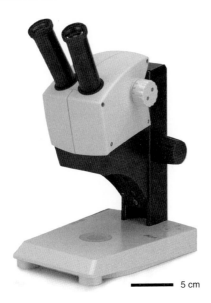

5 cm

Figure 6.2   The stereomicroscope, or dissecting scope, is used for examining microscopic or minute detail on the surface of leaves, fruits, or seeds, or to inspect physical evidence for botanical trace evidence. The Leica EZ4 is an education-grade microscope. Photo courtesy of Leica Microsystems, Inc., Bannockburn, Illinois, USA.

The stereo microscope (also known as a dissecting microscope)

A stereomicroscope, although seemingly a single microscope from the outside, actually consists of two microscopes that focus on the same point from slightly different angles to produce a three-dimensional image of the object being viewed (Figure 6.2). This microscope is used to examine the detail on small, yet still macroscopic, objects such as seeds, leaf fragments, or swatches of fabric for the presence of minute trace evidence such as trichomes, fibers, pollen, or filamentous or macroalgae (Figures 6.3 and 6.4).

Stereomicroscopes typically provide the user with a range of possible magnifications: most stereomicroscopes will offer 7–40X magnification, and higher-end models may obtain upwards of 60–100X magnification. *Depth of field* (an important parameter in microscopy and photography alike) is the vertical range about a subject under the microscope that is adequately sharp and focused for meaningful observation and discerning of structures. Although any given lens can precisely focus only at one distance (i.e., its *focal plane*), the decrease in sharpness is more or less gradual above and below the focal plane. Depth of field is greater at lower magnifications than higher magnifications. For example, at 40X magnification, the observer of a small flower may only see part of the flower in focus (say, the anthers), whereas at 10X the observer may perceive all parts of the flower to be in focus. To discern structures on a sample that lies outside of the depth of field, the observer must adjust the focus knobs.

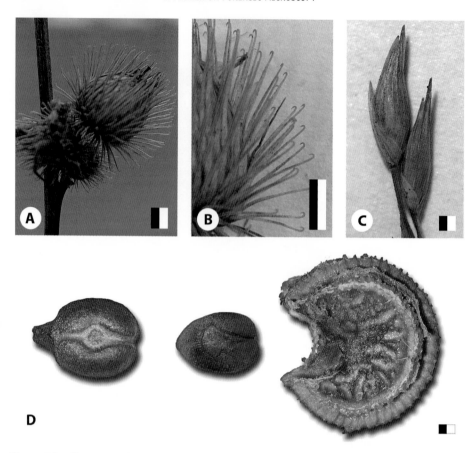

Figure 6.3 The stereomicroscope is used to examine the detail on small, yet still macroscopic, objects such as seeds, fruits, and flowers. Scale bars = 5 mm in A and B, 1 mm in C and D. A, B: Small fruits or dispersal structures such as those of the common burdock (*Arctium minus*) often have small hooks or barbs on them, allowing them to be easily picked up from a crime scene by unsuspecting passers-by. C: Small flowers, such as those of panic-grass (*Panicum virgatum*, switchgrass, is pictured here), often break away from the stalk easily and are picked up among the laces of shoes and cuffs of pants legs. D: Small seeds also may be picked up among shoe laces and cuffs of pants legs, may be found in tire treads, or they may indicate the components of a last "meal" when found in the stomach or stool of deceased accident or crime victims. Here, the seed of the common edible fox grape (*Vitis labrusca*, left) is pictured alongside the seeds of otherwise very similar, yet poisonous, berries of Virginia creeper (*Parthenocissus quinquefolia*, middle) and moonseed (*Menispermum canadense*, right). (Photograph of fox grape seed courtesy of Daniel J. Yoder.)

Unlike the compound microscope discussed later, the image produced with a stereomicroscope is upright and the object is usually illuminated from above (i.e., light is reflected off the subject) and there is no special preparation of the object needed. This is perhaps the most common microscope employed in forensic botany and will be indispensible in any laboratory regularly conducting forensic investigations (i.e., a

Figure 6.4 Various plant species have distinctive hairs, or trichomes, on their leaves, and the stereomicroscope is useful for examining these. Scale bars = 1 cm in A and C, 100 microns in B and D. The leaves of the leatherleaf viburnum, *Viburnum rhytidophyllum* (A), for example, have distinctive star-shaped trichomes (B), which may be rubbed off onto the clothing of unsuspecting passers-by (C). Examination of a patch of yellow "dust" on the sweater in C with the stereomicroscope revealed the star-shaped trichomes of the leatherleaf viburnum amongst the cotton fibers (D). (Please refer to the colour plate section.)

police forensics unit, a coroner's laboratory, or a forensic botanist's laboratory). Lower-end, yet adequate stereomicroscopes may cost several hundred US dollars (typically "education-grade" for high school and university classrooms) and higher-end scopes will cost several thousand US dollars.

The compound microscope

Compound microscopes (e.g., Figure 6.5) are higher-power microscopes in which light originating from a lamp in the base passes up through a sufficiently thin specimen mounted on a glass microscope slide (Figure 6.6) and through multiple lenses (hence, "compound" microscope) before reaching the eyes of the operator. The magnifications obtained with compound microscopes typically range from 40X to 600X or 1000X and are a product of the magnifications of the objective lens and the ocular (eyepiece) lens. For example, a specimen viewed with a 40X objective and the typical 10X ocular will be

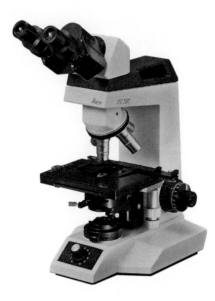

Figure 6.5   The compound microscope is used for examining microscopic detail on fiber or cellular/ tissue-level detail of plant structures thin enough to let light pass through them. The Leica ATC- 2000 is an education-grade microscope. Photo courtesy of Leica Microsystems, Inc., Bannockburn, Illinois, USA. (Please refer to the colour plate section.)

magnified 400X. As with the stereomicroscope, depth of field is greater at lower magnifications than at higher magnifications. To discern structures on a sample that lies outside of the depth of field, the observer must adjust the focus knobs.

Specimens for observation may be mounted if they are sufficiently thin to allow light to pass through them (e.g., microscopic algae, pollen, spores; Figures 6.7A–D) or they may have to be cleared (i.e., made clear with bleach-like solvents) or sectioned (i.e., a thin slice removed; Figures 6.7E–G) before mounting. Such microscopes will typically be available in the laboratory of a coroner and forensic botanist, but may not be part of the standard equipment in the laboratory at most police departments. Lower-end, yet adequate compound microscopes may cost around US$1000 (typically "education-grade"), whereas higher-end scopes will cost several thousands of dollars.

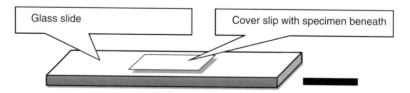

Figure 6.6   A diagram of a glass microscope slide, with specimen mounted beneath a glass cover slip. Scale bar = 1 cm.

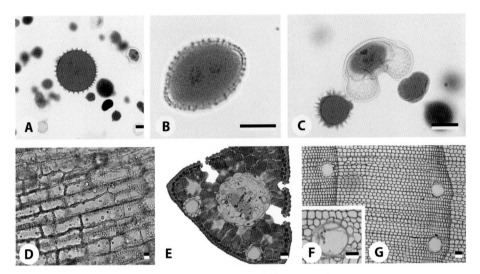

Figure 6.7   Examples of specimens typically examined with the compound microscope. Scale bars = 10 microns in A–C, 50 microns in D–G. A–C: Various pollen grains stained in different ways. D: A wet mount of a whole leaf of Elodea, a common freshwater plant. E: A transverse section of a pine needle, stained with the histological stains safranin and astra blue. F and G: Cross-sections through pine wood, showing the included resin canals typical of conifers. (Please refer to the colour plate section.)

The scanning electron microscope

Scanning electron microscopes (SEMs; e.g., Figure 6.8) use electrons instead of light to image a sample. To do this, the specimen is placed inside a sealed chamber and its surface is scanned with a high-energy beam of electrons in a regular raster scan pattern. Raster scanning is where a beam of electrons moves slowly from left to right in the field of view, then blanks to allow the frame to move slightly vertically, and then rapidly moves back to the left to scan another horizontal section of the frame. The energy of this primary electron beam energizes atoms at or near the specimen surface and causes secondary electrons to be emitted from the surface. These secondary electrons are then detected by the SEM's detector, where they convey information about the surface topography of the specimen, and a digital image is created on a monitor for later capture using film or digital imaging (Figure 6.9). Electrons have finer resolving capability than photons, and so much smaller and more closely spaced objects and particles can be clearly resolved with SEMs than with microscopes that use light. This also means that SEMs have the broadest range of magnifications possible among the microscopes, ranging from just 30X (comparable to that of the stereomicroscope) to 40,000X or more.

Specimens for observation with conventional SEMs are typically small: maximally 5–10 mm tall, 5–15 mm wide. The specimen chamber of a conventional SEM is under high vacuum and this requires that specimens for viewing be dry to avoid boiling off their water and subsequent specimen distortion. For fresh plant matter such as a

Figure 6.8 A scanning electron microscope has a rather large footprint compared to the other microscopes discussed. Pictured is an Amray (Bedford, Massachusetts, USA) 1820.

green leaf fragment or flower petal, the dehydration typically proceeds through a graded dehydration in solvents (e.g., alcohol, acetone) followed by removal of any remaining solvent in an instrument known as a *critical point dryer*, which dries the specimen without any artifact of the normal drying process such as wrinkling or

Figure 6.9 A scanning electron microscope was used to examine the surface detail of the distinctive yellow trichomes or hairs (B) on the stamens of this flower (A) of the Mexican spiderwort, *Tinantia pringlei*. Scale bars = 5 mm in A, 10 microns in B.

other distortion. Once dry, the specimen is mounted to an aluminum SEM "stub" with special glue or double-side tape. The specimen (with stub) is then given a thin coating of conductive metal (e.g., gold or gold-palladium) in an instrument known as a *sputter-coater* before fixing the stub with specimen to a grounded stage inside an SEM's specimen chamber for imaging. The conductive coating and grounding prevents the build-up of static electric charge on the specimen during irradiation with electrons: such charges would burn the specimen and interfere with its imaging. Non-conventional SEMs, called "environmental" SEMs (eSEMs), are specially engineered to work with the specimen under very low vacuum such that the specimen can be fresh (i.e., need not be dry). The gas inside the chamber of an eSEM also helps to dissipate the charging experienced by a specimen, such that coating the specimen in a conductive metal also is not necessary.

It is evident from the description above that SEMs may require more technical expertise to prepare specimens and to operate than the other light-based microscopes. SEMs also are typically much more expensive than the light microscopes discussed above, ranging from tens of thousands of US dollars to over US$100,000. Additionally, whereas SEMs are invaluable instruments for certain areas of scientific research, they will rarely offer much more to the forensic botanist than can be gleaned from a combination of a stereomicroscope and a compound microscope. As such, SEMs will not be and need not be found in a police department or coroner's facilities. Many universities and natural history museums, however, will possess SEMs. Even if a forensic botanist is based at such an institution, access to an SEM may be associated with specimen-processing and hourly operators' fees since such SEMs are sometimes components of larger microscopy facilities that seek to cover the costs of their operation, maintenance, and staffing.

The importance of reference collections in microscopic analysis

The fragmentary nature of most botanical trace evidence presents a challenge to the botanist attempting to confidently identify the species of the trace evidence in question. This is because botanists are generally trained or learn to identify plants using aspects of whole organs or even the whole organism. The taxonomic keys of books that botanists typically use to determine the identity of a plant to species also rely on the presence of some combination of whole flowers, whole leaves, etc. to work effectively. Thus, what is the forensic botanist to do when presented with a mere fragment of a dead leaf, when he or she was trained to distinguish the species of plants in a geographic area based on leaf shape and or flower form? Typically, an experienced forensic botanist can tentatively identify fragmentary plant evidence based on trichomes (hairs), veins, or other surface features, but will want to then compare that fragment to a more complete specimen of the species in a reference collection.

When the crime scene is known, the most important reference collection will be the collection of plants found at the crime scene. Ideally, the plants here are collected in prime condition, using a plant press or other appropriate collection vessel (e.g., a jar in the case of microscopic plants), and the specimens possess all the information necessary to confidently identify the plant species. Then the diagnostic minute or microscopic features of fragmentary trace evidence can be compared and possibly matched to those of the crime scene reference samples for positive identification.

The second important type of reference collection available to the forensic botanist is the herbarium. Herbaria are archived collections of dried, preserved plants housed in many universities and natural history museums. Such collections are not specific to a crime scene; rather, they are generally built by the cumulative efforts of students and researchers at these institutions over the years to document the plants they have encountered during field work for research projects or class projects. The ideal specimen has all the diagnostic features for a species, such as a portion of the shoot, flowers, or fruits, and comes with a label containing the identity of the species, the identity of the collector, and the exact location and date the plant was collected (found). Whereas herbaria at small universities may include collections of a few thousand plant specimens emphasizing the local or regional flora, herbaria at major institutions such as the New York Botanical Garden, Missouri Botanical Garden, and Kew Botanic Gardens house collections of several million specimens from many different parts of the world. Herbaria allow the botanist to confirm tentative identifications of trace evidence plant material, or even of unidentified reference samples collected from the crime scene. Furthermore, since herbaria typically contain multiple specimens of the same species and of closely related species, herbaria allow botanists to assess the extent of natural variability between plants of the same and separate species, thereby helping to determine the level of confidence that botanists can have in their conclusions.

Botanists who specialize in microscopic organisms such as diatoms or in the study of microscopic structures such as pollen will have made or have access to their own reference collections, consisting of microscopic slides or sometimes a printed photographic atlas of said organisms that are perfectly analogous to the herbarium specimen of a larger, vascular plant. Where the trace evidence in an investigation is a microscopic structure or organism such as a diatom or pollen grain, the trace evidence may not be fragmentary, yet the reference collection provides the forensic botanist with previously identified material to confirm his or her determination or provides him or her with some feel for the variability between cells or pollen grains of the same species.

Preparation and documentation of specimen evidence for the microscopic examination

Preparing for and documenting the microscopic analysis of plant material will generally depend upon the type of microscopy that will be employed. Accordingly, in this section, the reader will find a general overview of some methods used, organized by microscope employed. The section does not duplicate the more general, yet important, discussions found in other chapters of this book of issues in forensics such as chain of custody. Additionally, the forensic examination of certain microscopic botanical organisms or structures such as some algae, pollen, or spores is discussed at length in other chapters in this book. One concept important in all types of forensic microscopy that warrants mention here is that of *artifact*. Often, botanical trace evidence is recovered from a scene or physical evidence in less than ideal form, for example upon drying, the plant matter had wrinkled or been otherwise deformed from its natural appearance when living. These artifacts of drying will make comparison to reference samples (i.e., those which are either still fresh or at least were preserved properly to minimize artifact) more difficult. In order to bolster his or her tentative identification of such trace evidence, the forensic botanist may sometimes attempt to reproduce the artifact by exposing freshly sampled material (e.g., reference samples from a crime scene) to the same elements thought to have caused the artifact on the trace evidence sample.

Hand-Lens and stereomicroscope observation

Specimens for examination with a hand-lens and stereomicroscope do not need any special preparation. Already dry specimens are simply examined dry, notes taken, drawings and photographs taken, and then may be returned to their evidence bags. Fresh specimens such as green leaves, or fresh flowers or fleshy fruits can also be examined as-is, but it is especially important to document the morphology of these at the time of examination with drawings, notes, and photographs since they soon deteriorate with time. In order to preserve the diagnostic attributes of a fresh, wet specimen, it may be preferable to use solvent-based preservatives such as formalin or FAA (formalin:acetic acid:50% ethanol, 1:1:8 v/v; Johansen 1940), both of which will cause the loss of color, but will preserve the three-dimensional shape, form, and, in the case of FAA, internal cellular structures indefinitely. Such a preservation allows the material to easily be examined again for re-evaluation or exhibit during trial if the actual material, rather than photographs, is requested.

Compound microscope observation

Specimens for examination with a compound microscope will generally require some special processing before examination. First, a specimen sample must be thin enough to allow light to pass through it and small enough to be mounted to a glass microscope slide. Minute organisms or structures such as many algae, spores, pollen, and fibers are already thin enough and small enough to be mounted and examined nearly as-is. Wet mounts are where a pipette is used to place a drop or two of the medium (e.g., water or alcohol) containing the organism onto a glass slide and covered with a cover slip. Such wet mounts will eventually dry out; thus, beyond photographic documentation, permanent mounts (Johansen 1940) may have to be made for later reference.

Some aquatic plants have organs (e.g., leaves) so thin that they can be mounted whole onto a microscope slide (e.g., Figure 6.7D). However, the organs of most plants will need to be thin-sectioned prior to mounting to a slide for later examination with the compound microscope (e.g., Figures 6.7E–G). Free-hand sections of fresh or rehydrated tissue can be made with a razor blade. Finer observations of the anatomy of plant organs can be made with a microtome. To do this, the material (e.g., a leaf) is fixed in a preservative such as FAA, dehydrated through a graded ethanol to xylene solvent series, and then mounted into a square mold containing hot paraffin wax, which solidifies as a block (see Johansen (1940) and Hardy and Stevenson (2000) for examples). The wax block with embedded specimen is then fixed to a holder and sectioned using a microtome. After sectioning, the sections are mounted to a glass slide and typically stained to increase the contrast between cellular structures using standard histological techniques (Johansen 1940).

SEM observation

Specimens for examination with a scanning electron microscope will generally require some special processing before examination. Already dry specimens may be mounted to an aluminum SEM stub directly without the need for critical point drying. For small specimens, double-sided tape may be sufficient to secure the specimen, although glue may be needed for larger samples. It is important to know which face of the specimen you wish to view because, for all practical purposes, the gluing is permanent and the stuck surface will be forever lost to observation. For fresh specimens, if it is desired to examine them as close to their natural state as possible, without the distortive artifact of normal drying, critical point drying (CPD) will be necessary. In preparation for CPD, the fresh plant matter should be fixed in a fixative such FAA, then dehydrated through a graded ethanol series, and then transitioned to some transitional fluid such as acetone (although some users use ethanol for this). CPD can then proceed on the material using a critical point dryer. After CPD, the specimen is ready for mounting to the SEM stub. Once mounted to the stub, observation using a conventional SEM will require coating the specimen and part of the stub with a conductive metal such as gold or gold-palladium in a sputter coater. Afterwards, the material is ready for examination with the SEM.

References

Hardy, C.R. and Stevenson, D.W. (2000) Development of the flower, gametophytes, and floral vasculature in *Cochliostema odoratissimum* (Commelinaceae). *Botanical Journal of the Linnean Society*, 134(1), 131–157.

Johansen, D.A. (1940) *Plant Microtechnique*. McGraw-Hill, New York.

7 Plant anatomy

David W. Hall, Ph.D. and William Stern, Ph.D.

Plants are combinations of various structures such as roots, stems, branches, leaves, and flowers. Each of these structures is composed of various kinds of cells. Obviously, a tree has different kinds of cells in its stem than does a water-lily. The anatomy of plants can provide important forensic evidence. A plant anatomist (one who studies plant cells and tissues) can often determine the part of the plant to which certain cells belong. More commonly in forensics the various cells and tissues are used to try to determine the identification of the plant fragment. Plant fragments can be very difficult to identify, even for an anatomist. There are a few hundred thousand different kinds of plants in the world. Plants that are closely related can be expected to have very similar tissues and cells. Even if an anatomist can identify the fragment as being from a particular kind of plant, for example a cypress, there may be different species of cypress located in scattered areas of a region, continent, or several continents. The identification of a particular species of cypress could be crucial evidence and possibly place a vehicle or suspect at a particular scene. Identifying the plant as a cypress, even without the species, can provide information about the wet habitat in which it grows.

Closely related species may only differ in a single characteristic, which may not be present on the evidence collected. If a crucial anatomical character needed to determine a species is not present, the plant cannot be identified but still can be useful. It may be possible to determine a broader relationship such as a genus, family, or division, or, more simply, if the plant is woody or herbaceous. Such a relationship can be important to circumstantially link scenes or suspects. New DNA techniques may provide an alternative means of identification if definitive anatomical characteristics are not present.

During the growth of a plant, different cells and tissues develop at different times. The occurrence of a particular cell or tissue can show that a plant, even a very young seedling, has reached a certain stage of growth. Sometimes this can also enable a circumstantial match to similarly aged vegetation.

Woody plants can be used to help provide the relative time since death. Woody plants grow for more than one year and increase their total width with each succeeding year,

Forensic Botany: A Practical Guide, First Edition. David W. Hall and Jason H. Byrd.
© 2012 John Wiley & Sons, Ltd. Published 2012 by John Wiley & Sons, Ltd.

with stems, branches, and roots becoming wider. A central tissue (xylem), the wood, transports water and minerals from the roots to other parts of the plant. As the plant grows the outer cells (the sapwood) are active and the inner cells (the heartwood) are used for strength and become inactive. The inner cells are further strengthened by a compound called lignin, and provide the main support for the entire plant structure. In many woody species plant growth is very rapid in the early spring and slows gradually through the growing season, becoming almost non-existent during the cold months. When the plant is rapidly growing the plant produces very large cells to conduct water. As growth slows the water-conducting cells become smaller, until at the time of frost they are quite small. When looking at a cross-section of the wood, the contrast between the large cells of spring and the very small cells of winter is apparent. The cross-section of woody plants is circular, as is the ring produced by the differences in growth. In temperate climates the rings can be counted and equated with years of growth. A woody plant growing through a skeleton or on a grave can be said to be a certain number of years old. However, the plant may not have germinated immediately when growing conditions were suitable. The time interval indicated shows that the grave or skeleton would have been there at least that many years, but more time could be involved.

Not all woody plants, even in temperate climates, produce annual rings. Woody plants in warm or tropical environments that lack definite seasons sometimes do not produce annual rings. However, tropical climates can have wet and dry seasons, which can produce rings.

The irregularity of seasons in temperate climates can generate false rings. Exceedingly dry periods during the year and warm spells after a frost can be problematic. Additionally, several species normally present irregular rings. Some interpretations must be made by a dendrochronologist (one who studies tree rings).

Tree rings can also be used to show when damage occurred to a tree. Damage can be caused by a vehicle bumping into or running over a tree, a bullet shot into the trunk, or a body, soil, or other object placed on the root system or over a branch. For recent events the rate of leakage of the sap or the amount of sap sometimes can be used to determine relative time. If leakage of sap is to be helpful, it is important to have a specialist examine the scene as soon as possible. The specialist might need to utilize nearby vegetation for an experiment. The data will not be relevant if a longer wait has changed the scene. If considerable time has elapsed any rings that have developed since the damage can be counted.

If wood is used as a weapon, an anatomist might be able to identify the kind of plant from which the wood came by examining splinters or microscopic fragments. Sometimes the anatomist can match broken ends of the splinters or a damaged area to a tool mark. When a saw is used to cut wood the sawdust left in the teeth can sometimes be identified and subsequently matched. Sawdust left in a saw's teeth can be valuable in cases where high-priced timber has been cut or even in neighborhood disputes where local laws apply. Cells from plants can be helpful in cases of vaginal or anal penetration by wooden objects, or woody or herbaceous plants. In one case the few cells that remained were analysed to show that the plant was probably an incidental inclusion and not an intended weapon.

Timber can be extremely valuable and illegal harvesting is common. Logs or finished wood from illegal timber can sometimes be matched to the stumps. Illegal logging is frequent on land that is easily accessed, but not often visited by its owners, such as large blocks of land owned by governments, conservation organizations, and large timber companies. Rare timber in any area is vulnerable as the profits are large. Trafficking of some rare species is prohibited by various countries. Identification is crucial, as is the ability to find the approximate time the tree was felled. The tree could have been cut before it was protected by regulations. Matching tree rings from a piece of wood to known representative samples can determine the year it was cut and if it matches a particular stump. This technique can be also used to detect forgeries in furniture or art. Furniture forgeries can sometimes be revealed by showing that inexpensive wood was used in less visible parts of the construction. Also, if the art or furniture object is supposed to be representative of a certain era, the wood used to make the forgery might not have been available to the manufacturer during that era, therefore indicating a forgery.

Collection of samples for anatomical analysis at a crime scene follows the same guidelines as listed in Chapter 3. As with other plant evidence collected, if an anatomical examination is needed, consulting a forensic plant anatomist as soon as practical is recommended. If special collection techniques are required, they need to be determined by a forensic plant anatomist, who may need to visit the scene before changes to the vegetation can take place.

The Lindbergh case

The identification of wood to provide evidence in legal issues is rooted in the ability of investigators to understand the structure of wood and its relationship to the identity of the plant in question. Species of seed-bearing plants are so designated primarily on the characteristics of their reproductive and associated organs - flowers, fruits, cones - and not on the features of their vegetative structure, of which wood is a part. Thus, one should not expect to be able in all cases to identify the species of a plant solely from wood structure. In many instances, for example among the commercial timbers of North America, it is usually not possible to discriminate among the species of a genus. The so-called "southern pines", which is a group of species including slash pine, shortleaf pine, loblolly pine, pitch pine, longleaf pine, and others, is a good example of this. Native oaks can only be separated in red and white groupings based on the structure of the wood. The hard maples, including sugar maple, cannot be separated structurally from one another, but as a group they can be distinguished from soft maples, such as red maple and silver maple. Regardless of the constraint, wood structure and information on its growth and formation can be useful in establishing bases for evidence in courts of law regardless of issues of doubt about the identity of wood, some of which never reach the stage of litigation.

Probably the most celebrated instance of the use of wood as evidential material in a court of law, and which established, for all intents and purposes, the acceptance in the United States of scientific evidence in court, was the case of *State of New Jersey vs. Bruno Richard Hauptmann* for the kidnapping and murder of the infant son of

Charles A. and Anne S. Morrow Lindbergh. The then extraordinary testimony of Arthur Koehler, the "expert on wood", provided a precedent for the introduction into the courtroom of technical material for presentation to a lay jury. Thus, there is generally no need today to convince the court of the usefulness and validity of scientific evidence where the introduction of such information is germane to the prosecution of a case. It might serve the purpose of readers to detail that part of the Bruno Hauptmann trial dealing with the evidence from wood, and especially the then unprecedented testimony of Arthur Koehler, as a prime example of the use of wood in the courtroom.

Both Charles and Anne Lindbergh were well known figures in the late 1920s and 1930s, the era involved in this discussion. Charles A. Lindbergh had flown solo in a single-engine monoplane from New York City to Paris on 20 and 21 May 1927 and was the darling of society in the United States and throughout the world. He was the hero of the time and was honored for his achievement by several governments. Anne Morrow came from a distinguished family. Her father had been US ambassador to Mexico, a US Senator for New Jersey, and a partner in the prestigious banking firm of JP Morgan. Her mother, Elizabeth Cutter, had been acting President of Smith College and had published several books. Anne Morrow was a published author in her own right. Thus, the kidnap and murder of their 21-month-old son caught the sympathies and captured the imaginations of parents over the world perhaps as no other kidnapping before had done. The bare details of the crime were these:

(1) The Lindberghs spent weekends on their large estate in rural Hopewell, New Jersey, near the town of Flemington.

(2) During the evening of 1 March 1932, their son, Charles A. Lindbergh, Jr., was abducted from his second-storey nursery while his parents and nursemaid were at home. The nursemaid discovered the child's disappearance.

(3) Clues on the spot were meager:

(a) A wooden ladder in a clump of bushes about 60 feet from the house. It was homemade, crude, and consisted of three 7-foot sections, apparently assembled on the spot.

(b) The ground under the nursery window bore two indentations, such as might have been made by the rails of a ladder, and the whitewashed wall directly beneath the nursery window through which entrance was gained was smeared with soil in two places, as were the upper ends of the ladder rails.

(c) Footprints made by stockinged feet started from below the window and ended where the ladder was found.

(d) In the soft earth beneath the window there was a 3/4-inch-wide carpenter's chisel.

(e) A ransom note demanding $50,000 for the baby's return was located on the radiator grill inside the window.

There were negotiations with the kidnapper for a period of several weeks and eventually payment of the demanded ransom was made, but the baby was never delivered. The boat on Long Island Sound, on which the baby was supposed to have been held, was never located. On 12 May 1932 the baby's body was found in a shallow grave about five miles from the Lindbergh home. Apparently, the baby was killed as the kidnapper descended the ladder when a rung broke, throwing them both to the ground.

Shortly after the kidnapping was discovered and the police informed, the Governor of New Jersey (A. Harry Moore) assigned the superintendent of the New Jersey State Police (Col. H. Norman Schwarzkopf) to take personal charge of the investigation and President Herbert Hoover directed the Federal Bureau of Investigation, with its diverse resources, to lend every assistance within its power to find the perpetrator of the crime. After the initial investigative flurry of activity, frantic pleas from the distraught parents, the ransom money having been delivered, and the baby found dead, the wave of public outcry dimmed. Other news - bank crashes, Franklin D. Roosevelt's election as President, New Deal Legislation, and trouble brewing in Europe - served to dilute the immediate intensity of public concern over the kidnapping. During this period, however, Arthur Koehler, who was to be described during court proceedings as the "expert on wood," was engaged by the state to examine whatever evidence might be found in the construction and materials of the ladder, and law enforcement officials continued their efforts to locate the kidnapper.

Bruno Richard Hauptmann, a 34-year-old carpenter of German birth living in the Bronx, New York, was captured in September 1934 and subsequently charged with the kidnapping. It is necessary to know in this regard that as part of Roosevelt's New Deal recovery act, the United States had gone off the gold standard, and holders of both coin and gold certificates were ordered to turn them in to the Federal Reserve Bank in exchange for greenbacks. Hauptmann "passed" a gold certificate while purchasing gasoline for his car in New York City. The station attendant became suspicious that the customer was a "gold-hoarder" and marked the license number of his car on the margin of the bill. That bill proved to be one of those bearing a serial number matching a certificate included among those in the ransom payoff money. It was a simple matter then to locate the owner of the car and passer of the gold certificate, and Hauptmann was apprehended near his home.

A search of Hauptmann's person and garage turned up thousands of dollars in gold certificates bearing the serial numbers of the ransom money. Thus, Hauptmann was indicted, and on 2 January 1935 his trial began in the Hunterdon County Courthouse in Flemington, New Jersey for the kidnapping of the Lindbergh baby. The State of New Jersey appointed its attorney general, David T. Wilentz, to present its case, and Edward J. Reilly and his associates of the New York bar were assigned the defense. A great number of witnesses, both lay and expert, were called by the state in support of its case, most notable being the handwriting experts who the state contended proven beyond any doubt that the ransom and other notes were written by Hauptmann's hand. Following this, they introduced circumstantial expert opinion evidence - evidence that the ladder found on the premises after the baby's disappearance had been made by Hauptmann.

Hauptman had purchased some of the lumber and other parts of it had been taken from the attic floor of his home.

Author Koehler was, at that time, on the staff of the US Forest Products Laboratory in Madison, Wisconsin. He was a forestry graduate of the University of Michigan and through postgraduate work at the University of Wisconsin had earned an MSc degree. His work at the laboratory involved the identification of thousands of wood specimens each year in addition to the performance of services in analysing the growth patterns of wood, assessing strength and other mechanical properties of wood, and recommending utilization of milling methods for wood. Thus, he was well-equipped by formal study and experience to tackle the jobs involved in pinning down the origins of the kidnap ladder. It is well to recall at this point that Koehler's investigations were done *prior* to the location and indictment of Bruno Hauptmann as the perpetrator of the kidnapping. Koehler's research produced the following physical evidence:

(1) Two pieces of wood were entered as exhibits: one, a side rail of the upper section of the ladder; the other, part of a board taken from the flooring in Hauptmann's attic. Koehler testified that the ladder rail had been part of a longer floor board pried away from the joists in Hauptmann's attic and sawed from it. He reached this conclusion for several reasons:

   (a) When the ladder rail was placed over the attic floor joists, the nail holes still present in the ladder rail matched in number, position, and shape the holes remaining in the joists.

   (b) When the two boards, the board end from the attic and ladder rail board, were placed end to end, the growth rings in each matched perfectly.

   (c) The numbers of rings in each board were identical.

   (d) The sawdust on the ceiling plaster beneath where the floor board had been cut was the same kind of wood as the floor board from the attic and the ladder rail.

   (e) The width of the saw cut, that is, the kerf, was between 35 and 37 one-thousandths of an inch, and Hauptmann had two saws in his tool chest of exactly that width.

(2) The edge of the ladder rail had been planed with a hand plane and the planer marks on the wood indicated that the blade of the plane bore a certain pattern of nicks. With the court's permission, Koehler clamped a vise to the judge's bench and using the plane from Hauptmann's tool box he proceeded to plane a new piece of wood. The resulting pattern on the new wood showed the same pattern of scorings as that on the edge of the ladder rail.

(3) An examination of the ladder rails other than those traced to Hauptmann's attic showed the marks of a machine planer on their flat and edge surfaces. Each planer shows distinctive marks, and Koehler knew, if he could locate the mill that had a

machine planer producing marks identical to those on the kidnap ladder, he could begin to trace the origin and further movements of the wood in the ladder, from mill to retailer.

(a) First Koehler determined that the wood was what he called "North Carolina Pine." Actually, that pine wood was what would be termed today one of the southern pines, that is, one of a series of pine species indistinguishable from one another based on wood structure. This broad identification warned him that the wood originally came from the pine trees growing in the southern United States.

(b) By examining the markings made by the machine planer, Koehler determined that the top and bottom cutter heads of the planer were equipped with eight knives, and that the edge cutter heads held six knives. He knew from his experience that only two factories in the East manufactured planers of that kind. He found that these two concerns had manufactured and sold more than 1500 planers to mills in the East and South. Through the mail and by telephone, Koehler learned that 25 of these mills had purchased machines of the type used to plane the boards in the ladder.

(c) Koehler traveled to each of these mills and carefully examined the markings made by their planers until he found one which left marks identical to those in the ladder rails. That mill was owned by M.G. and J.J. Dorn of McCormick, South Carolina.

(d) His next step was to find where shipments of lumber planed by the Dorn machine had been sent. Records of the mill showed that 45 carloads of 1 × 4 (the kind of lumber used for flooring and in the ladder side rails) had been shipped to 25 Eastern and Southern lumber yards between October 1929 (when the planer had been placed into operation) and 1 March 1932 (the date of the kidnapping). Each shipment was traced and the effort paid off in September 1933, a year before Hauptmann was arrested. 1 × 4 flooring bearing marks identical to those on the kidnap ladder rails was found to have been shipped to the Great National Millwork and Lumber Company in Bronx, New York. The surprise came after Hauptmann's arrest when it was found that he had worked for this company for several months during 1931 and 1932 and had purchased such lumber from them.

Attorney General Willentz had vowed to hang the ladder, as he said, "right around his (Hauptmann's) neck. One side of that ladder comes right from his attic (he exclaimed), put there with his own tools, and we will prove it to you."

Heretofore, there had been no precedent for the appearance in court of a scientist versed in the identification of wood and similar areas of knowledge about wood structure and growth. Although Koehler was qualified by the prosecutor as an "expert on wood," defense counsel, Frederick A. Pope, protested to the court that "there was no such animal known among men." However, the graphic testimony of Arthur Koehler and the care used by Attorney General Wilentz to coach Koehler on how to phrase his explanations of technical material to a lay jury evidently were sufficient to convince the jury of the credibility of Koehler's testimony.

Hauptmann was found guilty and sentenced to death in the electric chair of the New Jersey State Penitentiary, with which he was so dealt on 3 April 1936. His wife, Anna, who died aged 95, continued to fight for recognition of her husband's innocence and sought his exoneration until her death. The well-illustrated volume by Anthony Scaduto purports to show that Hauptmann was framed and the author's arguments are at times convincing.

Besides the ready acceptance of scientific evidence today in courts of law (e.g., the increasingly wide use of DNA evidence), this particular case gave rise in May 1934 to what is commonly called in the United States the Lindbergh Kidnapping Law. It stipulates that when in the commission of a kidnapping a state boundary is crossed, the act is considered a federal crime and the death penalty is prescribed if the victim is not released unharmed.

Further reading

Anonymous (1935) Wood's case against Hauptmann. *American Forest*, 41, 209–211.

Busch, F.S. (1952) *Prisoners at the Bar - an Acount of the Trials of the William Haywood Case, The Sacco-Vanzetti Case, The Loeb-Leopold Case, The Bruno Hauptmann Case*. Section IV - "The Trial of Bruno Hauptmann for the Murder of Charles Lindbergh, Jr." Bobbs-Merrill, Indianapolis and New York, pp 201-208.

Cutler, D.F., Botha, T., and Stevenson, D.W. (2008) *Plant Anatomy An Applied Approach*. Blackwell Publishing, Malden, MA.

Evert, R.F. (2006) *Essau's Plant Anatomy*, 3rd edn. John Wiley & Sons, Hoboken, New Jersey.

Forest Products Laboratory (1935) *The Ladder, the Lumber, and the Laboratory*. Madison, WI.

Gummere, C.S. (1936) The state of New Jersey, defendant in error, v. Bruno Richard Hauptmann, plaintiff in error, Vol. 30. Reports of cases argued and determined in the Supreme Court of the State of New Jersey. Soney & Sage. Newark, pp 412-446.

Koehler, A. (1935) Who made that ladder? *Saturday Evening Post*, 207(42), 10–11, 84, 86, 89, 91-92.

Koehler, A. (1937) Technique used in tracing the Lindbergh kidnapping ladder. *Journal of Criminal Law and Criminology*, 27, 712–724.

Scaduto, A. (1976) Scapegoat: The lonesome death of Bruno Richard Hauptmann. G.P. Putnam's Sons. New York.

Stern, W.L. (1988) Wood in the courtroom. *World of Wood*, 41(9), 6–9.

8 Palynology, pollen, and spores, partners in crime: what, why, and how

Anna Sandiford, Ph.D.

Terminology

Forensic science is divided into several categories; the category into which palynology falls is "trace." Trace material includes microscopic particles of material that may be transferred at the time an event of interest occurs (Caddy 2001), for example the transfer of glass fragments to clothing at the time a window is broken.

The term "palynology" was originally used to refer to the study of pollen and spores (Traverse 2008) and people studying pollen and spores are still referred to as palynologists. However, many palynologists now use the term to include the study of all organic micro-organisms and material that remain in a sediment or sample after subjection to a specific series of chemical processes.

The term "pollen" is singular and plural; it is not generally considered correct to refer to "pollens." The term "pollen grain" is often used to refer to a single particle, or grain, of pollen. Spores can be included in the term "pollen."

What are pollen and spores?

Pollen is produced by seed-producing plants and plays a role in plant reproduction (see Chapter 2). Pollen grains are the male sperm cells that fertilize the female egg cells. The pollen of many plants is contained in flowers whereas some plants, such as pine trees, have pollen-producing cones instead of flowers. Other plants, such as ferns, fern allies, mosses, liverworts, and lichens, use spores in their reproductive process instead of pollen.

Forensic Botany: A Practical Guide, First Edition. David W. Hall and Jason H. Byrd.
© 2012 John Wiley & Sons, Ltd. Published 2012 by John Wiley & Sons, Ltd.

Most plants release pollen or spores at certain times of the year. Reproduction of plants generally occurs on the basis that if sufficient reproductive bodies are released, some of those reproductive bodies will go on to produce new plants. Pollen and spores that do not become new plants remain, unused, in the environment. It is this prevalence in the environment and their ability to be transported that makes pollen useful in both the forensic and non-forensic scientific contexts. The abundance of pollen in the modern-day environment is underlined by the vast numbers of people who suffer from hay fever, because it is usually pollen that causes the human body's hay fever reaction.

Chemical and physical resistance

The outer walls of pollen and spores are composed of organic compounds that are extremely resistant to many types of natural chemical and physical processes, which means they last a long time in the environment. Their resistance is apparent in their appearance in rocks millions of years old (e.g., Marshall and Fletcher 2002).

Pollen and spores are extracted from samples of material (such as soil and mud) using chemical processes that dissolve away most other material. The purpose of the part of the pollen grain that remains after the chemical processing (the outer wall) is to keep the reproductive part of the plant safe, which helps increase the chances of the plant successfully reproducing. The parts of the pollen grains that remain after chemical processing are therefore essentially extremely chemically resistant protective shells.

Other micro-organisms that are chemically resistant may also remain at the end of the pollen chemical processing procedure. These include parts of fungi, various types of algae (such as dinoflagellates, *Botryococcus*, and *Pediastrum*), and the interior coating of some foraminifera (a group of largely marine unicellular protozoan micro-organisms, informally referred to as "forams") as well as micro-organisms that are now extinct or whose predominant occurrence is in the fossil record, such as chitinozoans, acritarchs, and scolecodonts.

Samples that have been chemically processed for pollen may also contain fragments of resistant plant material, amorphous organic material, or fragments of charcoal (from fires, both natural and non-natural).

Although bacteria and fungi can damage and/or infest pollen grains and spores under certain environmental conditions, pollen and spores are notably more resistant than other types of organic material to degradation by micro-organisms.

Alternate wetting and drying of sediments containing pollen and spores will often accelerate damage to the grains. However, pollen can usually be found in most environments and preservation is particularly good in oxygen-poor (anoxic) environments. Natural oxidizing environments, such as high-energy rivers, and oxidizing chemicals, such as bleach, therefore tend to be the very damaging to pollen.

Where are they found and how do they travel?

Pollen is found in virtually every modern-day environment. This includes indoors, outdoors, and in water. In commercial and industrial environments, for example, some air-conditioning and air-quality units are specially designed to filter out microscopic debris from the air supply, including pollen.

The reason pollen is so widespread is a function of its reproductive purpose. Plants reproduce in a variety of ways and these mechanisms affect the design and numbers of pollen grains that are produced by the many different species of plants worldwide. The most common methods of dispersal of pollen and spores are to be blown by the wind, to be moved between individual flowers and plants by creatures such as insects and birds, and by water.

Wind

In order to ensure reproductive success, many plant species rely on sheer numbers of reproductive bodies. For example, a single pine tree can produce litres of pollen consisting of millions of individual pollen grains (Traverse 2008). This pollen is released into the air and the tree relies on wind currents transporting the pollen to cones on other trees, where the pollen may find its way into the cones to fertilize the female cells.

Pollen from plants that are wind-pollinated can be distributed over enormous areas, in the order of many tens of square miles, potentially crossing oceans (Close *et al.* 1978). When pollen grains are released from wind-pollinated plants, the pollen grains may be so abundant that they look like dust clouds. At certain times of year, surfaces such as the bodywork of cars may be coated with a thin layer of pollen "dust."

Pollen from such plants can be found in the water of deep oceans, thousands of miles from land and the source plant. This long-distance transport is, in part, a function of the shape and structure of the pollen grains: a central area containing the reproductive material to which are attached balloon-like sacs (Figure 8.1E and F). The sacs contain air and are thought to have adapted to keep the pollen grain buoyant for long distances as well as possibly help protect against the effects of desiccation, which may compromise the pollen's reproductive ability (Cranwell 1940; Jarzen and Nichols 1996).

Although not all wind-pollinated species have developed pollen grains with similar structures to those of the pine, there are many other wind-pollinated plants that use mass production of pollen grains as the basis of reproduction. These include rush-type plants (Restionaceae), sedges (Cyperaceae), and grasses (Poaceae).

Insects and birds

Some plant species rely less on volume of reproductive bodies and more on a direct method of transport of pollen from one plant to another. Such methods of transport may involve birds and insects such as butterflies, bees, and beetles.

Insect- and bird-pollinated plants usually produce much less pollen than wind-pollinated plants. The pollen is mostly distributed over a smaller area than wind-pollinated plants, in the order of square metres rather than square miles. These types of plants typically rely on insects and birds coming to them to collect the pollen and then transferring it to another plant of the same species. These plants often develop means of attracting the required pollen transporters, such as producing large, brightly coloured, scented flowers, for example daisies and roses.

Water distribution

Wetland plant species often rely on water to distribute their reproductive bodies. This means that the pollen or spores are usually restricted to the water environment. They may be abundant in the water environment but are far less common away from the water bodies in which the source plants live. Plants that fall into this category include water-loving mosses such as peat moss (*Sphagnum* spp.) and fern allies such as the quillworts (*Isoetes* spp.).

What does pollen look like?

Pollen grains are very small and are therefore examined using a microscope. The unit of measurement for pollen is the micron, one micron being 1/1000 of a millimetre or 0.001 mm, symbol µ. The term micron is another name for micrometre, the symbol for which is µm. Although micrometre is the more technically correct term (as it is one of the official SI measurements), micron is still used widely in the study of pollen.

Pollen grains vary in diameter from as small as 5 microns up to approximately 120 microns, which are sizes that are generally not visible to the naked eye. Some geological specimens can be much larger than this, in excess of 200 microns (Traverse 2008).

Pollen is extremely varied in its size, shape, and surface patterning. These are the features of pollen grains that are used to differentiate pollen from different plants and some pollen grains are more distinctive than others. For example, it is very difficult to distinguish between certain species of grass and for this reason the pollen is usually

→

Figure 8.1   Examples of pollen grains to demonstrate the variety of shapes and surface patterns that are used to distinguish them. All are shown at approximately 400X, except E, which is approximately 100X. A: Pollen of the small, common daisy, *Bellis perennis*. B: *Reevsiapolis reticulates*, pollen from an extinct species. The pink colouration is caused by the sample having been stained in order to enhance the surface patterns. This specimen is approximately 1.15 million years old. C: Pollen of *Fuchsia*, which is a common garden ornamental plant. D: Pollen grain from a tall forest gymnosperm tree (produces cones), *Prumnopitys taxifolia*, known in its native New Zealand as matai. E: The highly distinctive tri-saccate pollen of kahikatea, *Dacrycarpus dacrydioides*. The pink coloration is caused by the sample having been stained in order to enhance the surface patterns. This specimen is approximately 1.2 million years old.

Erratum Slip

Forensic Botany: A Practical Guide
Editors David W. Hall and Jason H. Byrd
HB ISBN 9780470664094, PB ISBN 9780470661239

Pages 130 and 131, Figure 8.1 should read as follow:

continued overleaf…

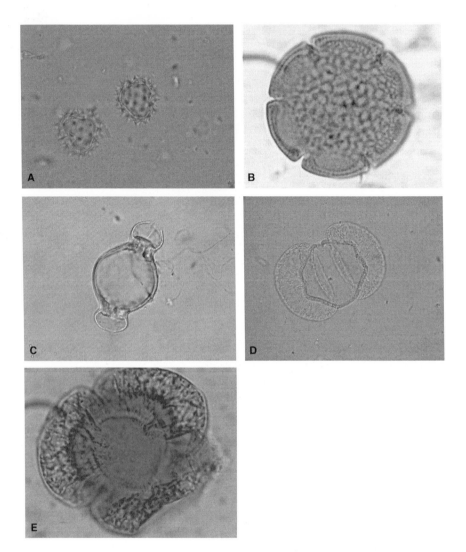

Figure 8.1 Examples of pollen grains to demonstrate the variety of shapes and surface patterns that are used to distinguish them. All are shown at approximately 400X, except E, which is approximately 100X. A: Pollen of the small, common daisy, *Bellis perennis*. B: *Reevsiapolis reticulates*, pollen from an extinct species. The pink colouration is caused by the sample having been stained in order to enhance the surface patterns. This specimen is approximately 1.15 million years old. C: Pollen of *Fuchsia*, which is a common garden ornamental plant. D: Pollen grain from a tall forest gymnosperm tree (produces cones), *Prumnopitys taxifolia*, known in its native New Zealand as matai. E: The highly distinctive trisaccate pollen of kahikatea, *Dacrycarpus dacrydioides*. The pink coloration is caused by the sample having been stained in order to enhance the surface patterns. This specimen is approximately 1.2 million years old. (The full colour version of this figure is available online.)

identified only to the family name, Poaceae. Plants such as oak trees or tree ferns can often be identified to genus level (*Quercus* and *Cyathea*, respectively). Yet other pollen grains can be identified to species, such as kahikatea (*Dacrycarpus dacrydioides*), a native tree of New Zealand that has very distinctive pollen (Figure 8.1F). Other examples of pollen grains are shown in Figure 8.1.

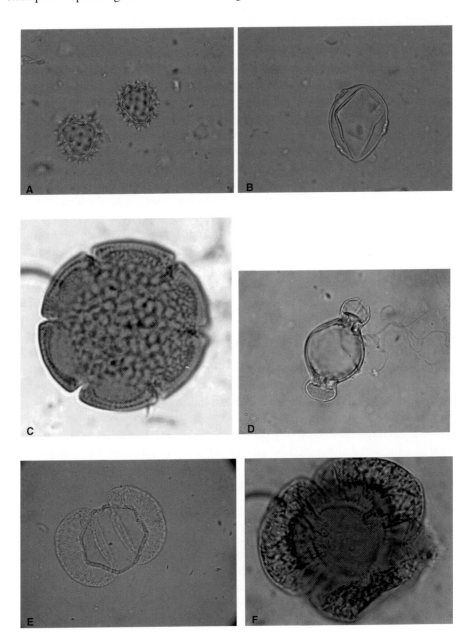

The use of pollen for non-forensic work

The study of pollen for non-forensic purposes is varied and ranges from the geological timescale through to archaeological applications as well as melissopalynology (the study of honey, including any pollen therein) and other food applications to determine origin of the source plants.

In geological applications, by examining the pollen assemblages, or "pollen signatures," of samples through a sequence of rocks and sediments, it is possible to see how the plant assemblages changed over time. This application of pollen analysis has helped determine climate change on the geological timescale of millions of years (e.g. Mildenhall 1980) through to thousands of years (e.g. Sandiford *et al.* 2003). It has also been possible to see how past volcanic eruptions affected vegetation (Sandiford *et al.* 2001), which, in turn, can help provide an indication of the effects of future eruptions. Archaeological applications are helping to unravel mysteries such as the timing of human arrival in isolated regions (e.g. McGlone and Wilmshurst 1999).

The application of pollen analysis to such topics is heavily reliant on the scientist's knowledge of how plants grow and reproduce, where they prefer to live in the natural environment, and the climatic conditions they require (see Chapter 2). Knowing what plants produce which pollen grains means that palynologists can use the pollen grains in a sample of soil to create a "virtual" picture of what plants must have been present in and around the area where the sample originated. Such a virtual picture will not be completely representative of the living vegetation at that site but it will provide a good indication of the general vegetation and environmental conditions in the local and wider area.

The use of pollen in the forensic setting

Forensic palynology uses the same principles and ideas as non-forensic palynology: identifying the pollen present in a sample gives an indication of the types of plants that were growing together which, in turn, gives an indication of the type of environment from which the sample originated.

When a forensic palynologist interprets his or her results there are several factors that should be considered although they will not all apply in all cases. As with any other casework, the individual circumstances of a case will dictate what is important. The factors to consider include the following.

Plants growing in the relevant location(s)

If pollen samples are collected from a crime scene, it is important for a record to be taken of the plants growing in and around that crime scene. This allows the forensic palynologist to relate the pollen in the sample to the vegetation at the crime scene.

Some plants produce very little pollen and might not be represented in control samples taken from a crime scene. Pollen grains that do not go on to fertilize a female are also subject to decay once they are deposited in the environment (e.g. Phuphumirat *et al.*2009). It is possible for a pollen type that does not occur in a sample taken from a scene to be identified in a sample taken from a suspect. Reference to the plants growing at the scene may allow a link to be made between a scene and a suspect that would otherwise not have been possible using the pollen sample alone.

Pollen rain

The amount and type of pollen found in a given location is referred to as the "pollen rain." This is a representation of pollen contributed by plants that were flowering at a given time over a given period. Pollen rain varies greatly, even over relatively short distances, and much work has been published worldwide that specifically assesses how the vegetation in a given area is represented in the pollen rain (e.g. Adam and Mehringer 1975, Olivera *et al.* 2009). Pollen rain does not necessarily represent accurately the vegetation of an area, that is, if 20% of plants in an area are grasses, it does not follow that 20% of the pollen rain will consist of grass pollen. Consideration needs to be given to the nature of the pollen rain and the plants that may have contributed to it. When interpreting pollen results consideration should also be given to the aerodynamics of a given location – pollen released from a plant in a windy location can be blown over much larger areas than pollen released from a plant living in a sheltered position.

Method of pollination

It is important to know the method of pollination of the plants that are represented in casework pollen samples. For example, because the pollen of wind-pollinated plants can be spread over a wide area, it may not be very important if it is found in a sample of soil because the same pollen grains might be present in samples taken from across a wide geographical area. However, it could be very significant to find pollen from plants that do not distribute their pollen very far from the source plant, such as pollen from plants that are insect or water pollinated.

Transfer and persistence

Transfer of pollen onto items will depend on the nature of the contact between the item and the pollen and the length of time over which the contact took place. Depending upon the pollen type, it may be necessary for an item to have sustained physical contact with a plant/soil sample/surface for transfer of pollen to occur. Simply walking close to some plants does not mean pollen will be transferred to clothing; it may require a person to brush against the plants in order for multiple grains of pollen to be transferred (Mildenhall 2006).

Once pollen has been transferred to an item, consideration has to be given to how long the pollen will remain on the item in question. This will include the type of item (clothing, spade, etc.) and activity of the item after transfer of pollen (vigorous activity, static, being washed, etc.).

Transfer and persistence issues are well recognized in relation to other types of trace material, such as glass and fibres (e.g. Robertson and Grieve 1999; Caddy 2001) and similar issues have been found to relate to pollen (Bull *et al.* 2006).

Types of sample(s) and location(s) from where they were obtained

Samples should be collected from a crime scene in order to provide a representation of the pollen of that location. These are usually referred to as control samples or reference samples. Samples from outdoor crime scenes are commonly soil, leaf litter, plant samples, and moss, the latter being particularly good at trapping and retaining pollen.

Samples need to be collected from locations of interest with care and with an appreciation for the nature of palynology. Unless there is information to the contrary, it would not normally be necessary to dig below the surface of an alleged crime scene in order to collect a pollen sample. An obvious exception to this would be when a body is recovered from a grave.

In relation to collection of samples from items found away from a crime scene or location of interest, such samples may be small in volume. For example, scuff marks on the knee region of a pair of trousers usually contain a smaller pollen assemblage (fewer pollen grains representing a smaller number of plant species) than a soil sample from a garden.

When should pollen samples be collected?

Pollen sample collection should be considered in a wide variety of cases, including drug trafficking and importation (particularly cannabis, heroin, and cocaine, which are directly derived from plants) and cases where there may be a link between people, items, and places (ranging from murder through burglary).

Analysing samples in drug cases can help determine where the drugs originated and where they may have been "cut" (e.g. Stanley 1992). Examples of cases where pollen has been used to provide a link between people, items, and places are well published, for example deer velvet theft (Mildenhall 1988).

How to collect and store pollen samples

Consideration of other types of scientific analyses

If a sample is specifically for pollen analysis then it can be collected and sent to the pollen laboratory directly. However, if a sample may need to be used for other types of scientific analysis then consideration has to be given to the order in which processing should take place. This is because pollen analysis is a destructive technique, which means that once the sample has been processed, everything else in that sample will be destroyed, apart from the pollen and other chemically resistant microparticles.

For example, a bloodied, grass-stained shirt may need to be examined for blood pattern analysis as well as pollen. It is usual to cut out a grass-stained area of material for pollen analysis, but this is a destructive technique and could disturb the blood pattern. In such circumstances it would be necessary for the blood pattern analysis work to be completed prior to the shirt being examined for pollen. Communication between the scientists and law enforcement agencies is therefore critical to ensure the maximum amount of information is obtained from the grass-stained, bloodied shirt.

Collection tools

Because pollen is so widespread, pollen samples can be collected from many types of material and surfaces, and most of the collection techniques are very simple. The basic tools of pollen sample collection are clean containers (such as 15-ml polypropylene tubes and small griplock bags), clean sampling tools (spatulas, teaspoons, and small brushes), labelling equipment (permanent marker pens, adhesive labels), and thin, disposable gloves (such as latex or non-latex single-use medical gloves). In the event that samples are collected at a scene at which there is no running water, another useful item of kit is alcohol wipes, which can be used to clean tools between samples. Tools should not be washed in water bodies (such as lakes, ponds, or rivers) as these will contain their own pollen load, which could contaminate samples.

All samples that could be used in forensic proceedings must be sealed into labeled, tamper-evident bags as soon as possible after collection, with each bag preferably having a unique number and/or barcode. Samples that are contained in non-secure bags are likely to be rejected for analysis because no guarantee can be provided that the samples were not contaminated after collection, whether accidentally or deliberately.

Accidental contamination

When dealing with very small quantities of material that often cannot be seen with the naked eye, precautions have to be taken to ensure that samples do not become accidentally contaminated. This means that the same sampling precautions and procedures should be observed for collection of samples for pollen analysis as are applied for other types of trace material such as fibres, glass, and paint (Caddy 2001; Robertson and Grieve 1999).

Accidental contamination could also occur if a sampling tube was not clean when it was used; dust, which often contains pollen, could introduce pollen grains into a sample that are unrelated to the case. If the forensic palynologist is not aware that the sample was contaminated then the contaminant particles would be included in the interpretation and could adversely affect the result.

There are other ways in which samples can become accidentally contaminated, such as using the same tool to collect two different samples without cleaning the tool in between taking each sample. It is therefore important that the people collecting the samples and/or who are present at the crime scene or in any item examination are aware of the risks of accidental contamination and take suitable precautions to minimize the contamination risk.

The scientists and/or technicians who will be processing the samples should be aware of potential laboratory-based contamination issues (Loublier 1998; Moore *et al.*1991) and should have appropriate procedures in place to minimize the risk.

Plants and soil

In many cases involving pollen, comparison of samples from different locations is undertaken. If, for example, it is suspected that a vehicle drove to a riverside location, samples need to be collected from the vehicle and also from the riverside location.

Pollen often adheres readily to the surfaces of plants, particularly hairy or sticky plant parts such as the leaves and flowering heads of cannabis plants. Collecting a sample from a batch of cannabis head material therefore just requires the plant material being selected and placed in a labelled, tamper-evident bag. Other drugs derived from plants such as cocaine can also be examined for the presence of pollen and there is no special collection technique; the drug itself can be submitted to the laboratory for processing (e.g. Stanley 1992). Laboratories that process drugs usually require a special licence that allows them to receive, store and handle illegal or restricted drugs.

Pollen is also very common in soil, dirt, and mud, and a sample of any of these can be collected from any surface on which it is present. For example, casework has involved scraping soil from the soles of shoes or collecting samples from the surface of a driveway using a new toothbrush. Soil or debris samples from crime scenes can be collected into sample tubes or clean griplock bags using a clean spoon.

Clothing and footwear

Entire clothing items can be collected and sent to the laboratory. It is possible to cut out sections of clothing on which soil smears or grass stains are present but discussion should take place with other scientists and the law enforcement investigators prior to this occurring (see earlier section). The cuffs of a sweater could be washed and the pollen collected from the washings. In these situations, the clothing item can be placed in a tamper-evident bag and sent to the laboratory for specialist examination. However, it is important for the pollen analyst to know how the clothing was handled prior to being packaged in the secure bag; placing clothes on a dirty floor prior to packing could result in accidental contamination. Similarly, shoes can be packaged and sent to the laboratory.

It must be remembered that unless samples from items such as clothing and shoes are collected prior to the items being sealed in evidence bags, soil and loose material may dislodge from their original locations on the shoes/clothing during transportation. If a pair of shoes is received at the laboratory and mud has shaken loose during transit, it may not be possible to indicate from where on the shoes (or even from which shoe, if packaged as a pair) the mud originated. Original location on a shoe may be important: mud on the top of a shoe may have been present for some time; mud on the sole of a shoe (rather than embedded in a tread) must relate to one of the most recent wearings of the shoe.

Vehicles

It is commonplace for vehicle seats and boot spaces to be tape-lifted in order to collect hairs and fibres from surfaces. This involves strips of clean, unused sticky tape being placed systematically over the relevant surface areas. The strips of tape are then usually stuck to transparent acetate sheets so they can be searched at the laboratory.

It is often considered that tape lifts can be used only for fibres or pollen, not both. However, new sampling and processing techniques are being developed that will allow tape lifts to be processed for both pollen and fibres without destruction of either.

Consideration should be given to interior surfaces of vehicles such as footwells, pedals, and mats. It is important to have an understanding of the period of time any such samples may represent. Case circumstances will therefore dictate whether or not there is value in collecting samples from these locations. Collection of fresh mud from the foot pedals of a brand new car is likely to have greater scientific significance than collecting a sample from the driver's floor mat of an old, well-used, dirty farm vehicle. Similar consideration also needs to be given to samples collected from the exterior of vehicles such as wheel arches, although there are cases where pollen (in conjunction with other soil examinations) from the exterior of a vehicle has helped direct a law enforcement investigation and locate the scene of a murder and burial (e.g. Brown 2000).

Human bodies

Collecting pollen samples from the human body can be more complicated than collection of other types of pollen samples. Collection from cadavers is best undertaken by a forensic pathologist; hair washings can be undertaken by mortuary assistants or scenes of crime officers who attend the autopsy. The hair of living subjects is easier to wash than that of the deceased, particularly if a corpse has begun to decompose and the scalp skin has started to slough off (Wiltshire 2006a).

Samples from beneath finger- and toenails are easily collected: the nails can either be clipped or the material can be scraped from beneath them. Smears of soil or similar material can be scraped from the surface or, in the case of a cadaver, the skin may be excised. Once again, consideration must be given to other types of scientific samples that may need to be recorded and/or collected, such as cellular material, blood, blood patterns or wound geometry.

In some cases, it may be appropriate to collect pollen from the nasal cavities of cadavers. This has often involved removal of the brain and washing of the nasal cavities from inside the skull (Wiltshire and Black 2006) but can also be undertaken by a forensic pathologist using a simple, non-invasive swab-and-tube sampling kit. Sampling for pollen from the nasal cavities using a swab-and-tube kit can be undertaken in such a way as to also allow a nasal swab to be taken for the presence of drugs.

Other items

Samples can be collected from a wide range of surfaces and objects, even after items of interest have been removed. One case involved examination of polymer sheeting in which cannabis resin had been wrapped at the time of a multi-tonne seizure. The bulk quantity of cannabis resin had already been destroyed so it was not possible to collect samples from within the blocks of cannabis resin, which would have been preferred. However, sub-millimetre sized fragments of resin were present in the folds of the polymer sheeting and these were collected into a sampling tube using tweezers and a hand-held lighting source. The total volume for analysis was less than half a gram but it was more than sufficient for the purposes of pollen analysis.

Household dust and dust/dirt on any surface such as carpets, inside suitcases, and mats will contain pollen. Collection of samples using a clean vacuum cleaner and new vacuum bag could be carried out in difficult locations, but it would be important to also submit to the laboratory unused vacuum bags from the same batch so that an assessment could be made of any pollen in the bags that was unrelated to the issue under investigation.

Any moderately sized object can be placed into an appropriately labeled exhibit bag, sealed, and sent to the laboratory for examination. This could include items like cloths, ropes, bricks, bats, tools; the list is extensive and only limited by individual case circumstances.

Water can be collected into secure clean jars (either polymer or glass). Up to 1 litre should be sufficient.

Storage of samples

As indicated previously, although pollen grains are resistant to many chemicals, alternate wetting and drying of samples and bacterial/fungal growth can damage pollen grains. The stability of moisture content and temperature is therefore important in order to preserve pollen grains. For this reason, pollen samples are generally kept in a fridge once they have been collected. However, unlike some DNA samples or tissue/body fluid samples, it is not usually necessary to transport pollen samples under refrigerated or dry ice conditions.

How many samples to collect?

The number of samples collected in a given case will depend on the case itself. In a cannabis case where the country of origin of the drug is being investigated, only one or two samples may be necessary. In a case involving a cadaver, a vehicle, and a crime scene, it may be necessary to collect several samples from each.

The number and type of sample to be collected can be discussed over the telephone/internet with a forensic palynologist or the forensic palynologist could attend the scene(s).

Who can collect pollen samples and where can an analyst be found?

Pollen analysis can be broken down into three basic stages: collection, processing, and identification/interpretation.

Collection

Anyone with a small amount of training can collect samples for pollen analysis, whether crime scene analysts, major case investigators, or forensic pathologists. Such training is simple to arrange and is usually conducted by forensic palynologists. However, as with any other form of material collected for forensic purposes, the quality and value of the final interpretation will be directly dependent on the quality of the initial sampling. It is therefore extremely important to ensure the sample is collected correctly then packaged, sealed, labelled, and stored appropriately.

In many situations, pollen sampling can take place after other samples have been collected and this can be organised by a suitably experienced crime scene manager.

If time and cost do not allow, it is not vital for a forensic palynologist to attend a crime scene, particularly if the crime scene is properly documented (see the section on costs and turnaround times, below).

Processing

Pollen processing should only be undertaken by appropriately trained staff who have an understanding of the nature of pollen and the issues of accidental contamination in the laboratory, and who have been trained in the handling of chemicals.

Identification of pollen and interpretation of findings

The identification of pollen and spores is a highly skilled occupation. A comprehensive knowledge of the world's general flora is required as well as an understanding of specific botanical issues, geographical locations, and geological and physical geographical processes.

Once all the pollen grains in a sample have been identified, it takes a forensic pollen expert to interpret the results. As previously discussed, many things have to be taken into account when interpreting pollen profiles, including methods of pollination, the period of time that each sample may represent, geographical variations, and specific case circumstances.

Costs and turnaround times

Although pollen identification and interpretation is a highly skilled job, the cost of sample analysis should not be prohibitive. Pollen sampling does not require expensive equipment. Pollen processing largely involves readily available chemicals and common laboratory equipment and glassware; the most expensive piece of equipment is probably a centrifuge of sufficient size to hold the processing tubes.

If crime scene attendees (such as crime scene investigators and police officers) are properly trained then there should be no need for a forensic palynologist to attend a crime scene on an emergency call-out basis. Although it is advantageous to visit a crime scene, if the scene is properly recorded through video, photographs, diagrams, and notes then the forensic palynologist's work should not be compromised if they do not actually attend the scene. Delaying post mortems for the attendance of a forensic palynologist is not ideal as this adds time and therefore cost to the investigation and also could delay release of the body to the family.

Depending on the laboratory facilities, the maximum manageable number of pollen samples for a scientist to process at any one time is about eight. Depending on the types of material that make up the samples, eight samples would take approximately one working day to process and mount on microscope slides, ready for examination.

Generally speaking, it should be possible to provide a pollen report involving around ten pollen samples within four weeks of the samples arriving at the laboratory.

Case examples

Forensic palynology is currently an under-utilized science, although there are many examples where pollen analysis has been extremely useful in forensic casework. These cases usually relate to serious charges like murder, but use in volume crime such as burglary is equally applicable – as long as there is sufficient material for collection then a sample can be analysed. Case examples are not always about the actual crime committed; sometimes, the circumstances dictate that the analytical work takes an unexpected turn.

Murder and genocide

Murder investigations are amongst the best publicized and there have been many examples where pollen analysis has assisted with murder investigations. A particular well-known English case is that of the murders of Holly Wells and Jessica Chapman, two schoolgirls who disappeared in the summer of 2002. Their bodies were found nearly two weeks after they disappeared, in a rural location. Pollen recovered from samples taken from the suspect's vehicle showed good correspondence to pollen from the outdoor crime scene (Wiltshire 2006b). The case also involved botanical findings relating to the broken stems of nettles and the speed at which new growth would have occurred after breakage.

One of the most high-profile applications of forensic palynology and murder relates to war crimes investigations. Pollen was used as part of investigations for the International Criminal Tribune for the former Yugoslavia to determine whether or not bodies in graves were killed elsewhere and transported to the grave from which they were exhumed. Some mass graves in north-east Bosnia that were used by perpetrators to bury victims of genocide were dug up again by the perpetrators, and the bodies removed and re-buried in other locations with fewer bodies in each grave. By doing this, it would appear that the deceased were not necessarily the victims of genocide but victims of smaller confrontations. War crimes investigators exhumed the smaller graves in 1997 and took soil samples from the clothes of the victims. The soil and pollen from the clothing of the exhumed bodies, along with other information, linked them to a mass grave in a different location, which, in turn, provided support for claims of genocide (Brown 2006).

Investigative work

Pollen analysis can be used as part of the intelligence-gathering process. The information obtained for intelligence purposes is not necessarily used for prosecuting a single case. Rather, it provides background information that can be used over time to help identify issues that are of interest and form the basis of a criminal investigation.

One example includes recovery of pollen from bulk quantities of drugs such as cannabis resin and their associated packaging to help determine in which country the source plants were grown. Cannabis plants collect pollen of other plants on their leaves and flowering heads. This pollen can be extracted from the leaves, flowering heads, resin, or oil, and analysed to identify the locations where the cannabis plants were grown, processed, and subsequently handled.

Intelligence agencies can use this sort of information to build up a picture of the path drugs consignments may take prior to seizure.

Accidental contamination

Interpretations in any case will only be as good as the information on which they are based. If the samples are compromised, the results may be invalidated. The key is being able to determine the reliability of the results. An example of accidental contamination compromising forensic palynological interpretation started with police raiding a property in the search for cannabis.

Two buckets of cannabis were located, one inside a shed and one under an old oil drum. The aim of the pollen analysis was to determine whether or not the two lots of cannabis had been grown at the property and/or if they had been grown during the same growing season.

As part of their work the forensic palynologists reviewed photographs the police took on the day that they entered the property and seized the cannabis. It was apparent that the police tipped the cannabis from under the oil drum onto the ground and took a photograph of it. The cannabis was then scraped back into the bucket, and the bucket carried over to a blue tarpaulin where it was tipped out again, spread out, and photographed. The cannabis was then scooped up and put back in the bucket.

The cannabis from the bucket under the oil drum was then poured onto the same blue tarpaulin. By pouring the second lot of cannabis onto the same sheet on which the first lot of cannabis had been poured, the police accidentally contaminated the second lot of cannabis. It was therefore not possible to determine whether the similarity between the pollen from the two sets of cannabis was the result of all the cannabis plants being grown at the same time (the prosecution's proposition) or if it had been grown at different times or possibly in different areas (the defence proposition). This caused a problem for the prosecution. In conjunction with other evidence heard at trial, the defendant was found not guilty.

Absence of pollen

The casework examples in this chapter have involved analysis of samples for the presence of pollen. In some cases, it is the lack of pollen that is significant. In this example, an allegation of rape resulted in the clothing and shoes of a suspect and a complainant being seized.

The suspect's shoes were examined and it was noted that they were not new and displayed evidence of the usual wear and tear associated with training shoes. The surfaces of shoes can be brushed and the debris analysed for pollen. However, in this case, the samples taken from the shoes of the suspect contained no pollen at all. This was an extremely unusual finding because even shoes that appear clean to the naked eye would be expected to yield some pollen grains, particularly if they are not new. The shoes were noted as being exceptionally clean and the white areas on the shoes were whiter than would be expected for their age. The lack of pollen suggested that the shoes had been thoroughly cleaned, possibly with bleach. This in itself was of interest to the police. Pollen was recovered from the suspect's clothing, and having considered all evidence presented he was found guilty at trial.

Summary

Forensic palynology is currently an under-utilized area of forensic science, although its application has been recognized for several decades (e.g. Bock and Norris 1996; Walsh and Horrocks 2008; Mildenhall 2009). Pollen is a valuable form of trace material that is easily collected by non-pollen experts as long as due consideration is given to contamination issues, which are the same types of issues that relate to other forms of trace material routinely collected for forensic casework.

Pollen sample processing and interpretation does not need to be expensive or time-consuming but it does need to be undertaken by suitably qualified individuals who have knowledge not only of palynology but also of forensics and the requirements of the courts. When used correctly, pollen analysis can be an extremely useful tool for investigative work as well as confirming or refuting alibis and proving or disproving links between people and case-specific places and items.

As with all areas of forensic science, it is not just the science that assists with casework. It is also critical to have an understanding of how samples were collected and where they fit into an overall case.

References

Adam, D. and Mehringer, P. (1975) Modern pollen surface samples – an analysis of subsamples. *Journal of Research of the US Geological Survey*, 3, 733–736.

Bock, J. and Norris, D. (1996) Forensic botany: an under-utilised resource. *Journal of Forensic Science*, 42, 364–367.

Brown, T. (2000) Going to ground. *Police Review (UK)*, 4 February, pp 18–20.

Brown, A. (2006) The use of forensic botany and geology in war crimes investigations in NE Bosnia. *Forensic Science International*, 163, 204–210.

Bull, P., Morgan, R., Sagovsky, A., and Hughes, G. (2006) The transfer and persistence of trace particulates: experimental studies using clothing fabrics. *Science & Justice*, 46, 185–195.

Caddy, B. (ed.) (2001) *Forensic Examination of Glass and Paint*. Taylor & Francis, London.

Close, R.C., Moar, N.T., Tomlinson, A.I., and Lowe, A.D. (1978) Aerial dispersal of biological material from Australia to New Zealand. *International Journal of Biometeorology*, 22, 1–19.

Cranwell, L.M. (1940) Pollen grains of the New Zealand conifers. *New Zealand Journal of Science and Technology*, 22B, 1–17.

Jarzen, D.M. and Nichols, D.J. (1996). Pollen. In: Jansonius, J. and McGregor, D.C. (eds), *Palynology: Principles and Applications, Volume 1: Principles*. American Association of Stratigraphic Palynologists Foundation, Dallas.

Loublier, Y. (1998) Evaluation of indoor passive pollen sedimentation over 1 year: a possible source of contamination? *Aerobiologia*, 14, 291–298.

Marshall, J. and Fletcher, T. (2002) Middle Devonian (Eifelian) spores from a fluvial dominated lake margin in the Orcadian Basin, Scotland. *Review of Palaeobotany and Palynology*, 118, 195–209.

Mcglone, M. and Wilmshurst, J. (1999) Dating initial Maori environmental impact in New Zealand. *Quaternary International*, 59, 5–16.

Mildenhall, D.C. (1980) New Zealand Late Cretaceous and Cenozoic plant biogeography: a contribution. *Palaeogeography, Palaeoclimatology, Palaeoecology*, 31, 197–233.

Mildenhall, D.C. (1988) Deer velvet and palynology: an example of the use of forensic palynology in New Zealand. *Tuatara*, 30, 1–11.

Mildenhall, D. (2006) *Hypericum* pollen determines the presence of burglars at the scene of a crime: an example of forensic palynology. *Forensic Science International*, 163, 231–235.

Mildenhall, D. (2009) Forensic palynology: an increasingly used tool in forensic science. *European Journal of Aerobiology and Environmental Medicine*, 2, 7–11.

Moore, P.D., Webb, J.A., and Collinson, M.E. (1991). *Pollen Analysis*, 2nd edn. Blackwell Scientific, Oxford.

Olivera, M., Duivenvoorden, J., and Hooghiemstra, H. (2009) Pollen rain and pollen representation across a forest-parano ecotone on nothern Ecuador. *Review of Palaeobotany and Palynology*, 157, 285–300.

Phuphumirat, W., Mildenhall, D., and Purintavaragul, C. (2009) Pollen deterioration in a tropical surface soil and its impact on forensic palynology. *The Open Journal of Forensic Science*, 2, 34–40.

Robertson, J. and Grieve, M. (1999) *Forensic Examination of Fibres*. Taylor & Francis, London.

Sandiford, A., Alloway, B., and Shane, P. (2001) A 28,000–6,600 cal yr record of local and distal volcanism preserved in a paleolake, Auckland, New Zealand. *New Zealand Journal of Geology and Geophysics*, 44(2), 323–336.

Sandiford, A., Newnham, R., Alloway, B., and Ogden, J. (2003) A 28,000-7,600 cal yr BP pollen record of vegetation and climate change from Pukaki Crater, northern New Zealand. *Palaegeography, Palaeoclimatology, Palaeoecology*, 201, 235–247.

Stanley, E.A. (1992) Application of palynology to establish the provenance and travel history of illicit drugs. *Microscope*, 40, 149–152.

Traverse, A. (2008). *Paleopalynology*, 2nd edn. Springer, Dordrecht.

Walsh, K. and Horrocks, M. (2008) Palynology: its position in the field of forensic science. *Journal of Forensic Science*, 53, 1053–1060.

Wiltshire, P. (2006a) Hair as a source of forensic evidence in murder investigations. *Forensic Science International*, 163, 241–248.

Wiltshire, P. (2006b) Consideration of some taphonomic variables of relevant to forensic palynological investigation in the United Kingdom. *Forensic Science International*, 163, 173–182.

Wiltshire, P. and Black, S. (2006) The cribriform approach to the retrieval of palynological evidence from the turbinates of murder victims. *Forensic Science International*, 163, 224–230.

9 Algae in forensic investigations

Christopher R. Hardy, Ph.D. and John R. Wallace, Ph.D.

Algae (singular, alga) and other aquatic plants are the primary producers in all water bodies (e.g., streams, lakes, estuaries, and oceans) and make up most of the biomass in these aquatic ecosystems. Aquatic ecologists have long recognized the fundamental importance of algae to the functioning of these ecosystems, as well as the role they play as determiners or as indicators of water quality within them. It is this sound ecological foundation that has lead scientists to explore their role in forensic investigations. Due to the prominent role and prevalence of algae in aquatic ecosystems, it is very likely that objects or persons involved in an accident or crime that takes place in water, even if only partially, will have algae in or on them. Particular types (species) of algae exist in certain terrestrial environments such as the shaded (e.g., the north side in the northern hemisphere) bark of trees, the shaded facades of houses, or in soil. It follows from the fact of their prevalence and importance in nature that algae have a high probability of proving useful to a forensic investigation in the hands of a capable botanist.

Finding an algal botanist and identifying algae

As discussed in Chapter 1, it is generally true that, given the paucity of specialists in "forensic botany" *per se*, professional botanists, known as plant systematists or ecologists, will generally have skills sufficiently broad to assist in forensic botany investigations. However, most systematists and ecologists specialize in higher (more advanced) plants of terrestrial habitats and may not have the skills or desire to efficiently conduct a forensic investigation that is likely to involve algae. Therefore, the search for an expert to help with algal identification may instead need to target biologists known as *algologists*, *phycologists* (phycology is the study of algae), *algal systematists*, *aquatic biologists*, or *freshwater* or *marine ecologists*. Such biologists are generally employed as researchers/educators at educational or research institutions such as universities or

Forensic Botany: A Practical Guide, First Edition. David W. Hall and Jason H. Byrd.
© 2012 John Wiley & Sons, Ltd. Published 2012 by John Wiley & Sons, Ltd.

natural history museums. Due to the prevalence of accidents or crimes in or around water, as well as the desire for timeliness in their investigation, police departments and other investigative agencies are strongly encouraged to identify willing and able biologists at nearby institutions *before* they are needed.

General information on the diversity and biology of algae can be found in the section on algal diversity below, and more details can be found in the texts by Graham and Wilcox (2000), Sze (1998), and Lee (1999). For the taxonomic identification of algae, one could start with the classic, well-illustrated, and practical reference by Palmer (1959). Even though Palmer's work is over 50 years old, the illustrations and keys are very good and are sufficient to identify many different groups of species. They will generally suffice to match the algal flora from a piece of evidence to a crime scene, for example. Furthermore, the Palmer work provides indications of the algal communities that are distinctive to certain types of water bodies (e.g., lakes, reservoirs) or water qualities (e.g., clean vs. polluted waters). More recent and detailed taxonomic keys are provided in Dillard (1999) and the American Water Works Association (2002) for freshwater algae of all types, Cox (1996) for freshwater diatoms, and Patrick and Reimer (1966) for fresh- or saltwater diatoms in the United States. Pollanen (1998) provides many pictures and descriptions of the major diatoms typically found in the analysis of putative drowning victims.

DNA-based methods for the identification of algae, which do not require taxonomic keys or morphological analysis, are beyond the scope of this chapter and the reader is referred to the chapter on DNA forensics (Chapter 5). The methods discussed in that chapter will generally apply to algae since the chloroplasts of algae and all plants are related, and the methods of DNA-based identification using the chloroplast genome are essentially the same across the species. As with any DNA analysis, it is extremely important to avoid contamination of samples with other algae since the polymerase chain reaction (PCR) method used to amplify DNA for subsequent examination is very sensitive.

Algal diversity

Algae come in many forms, sizes, even colors, and it is important to be at least remotely familiar with this diversity to ensure that their presence and potential utility in a forensic investigation is not missed.

Unicellular (single-celled) algae, although microscopic and therefore "invisible" to most investigators and perpetrators of crimes, are the most abundant and widespread of all aquatic organisms. Thus, it should always be assumed that algal evidence is present on an object or body recovered from water, or on any article of clothing thought to have been worn by a person suspected to have been involved in the accident or crime while in contact with the water at a scene – even if the algae cannot be seen with the naked-eye. Some of the more common examples of unicellular algae of freshwater (ponds, lakes, streams, and rivers) include freshwater species of diatoms, chlorella, chlamydomonas, euglenoids, and dinoflagellates, whereas some of the more common algae of saltwater include marine species of diatoms and dinoflagellates (Figure 9.1).

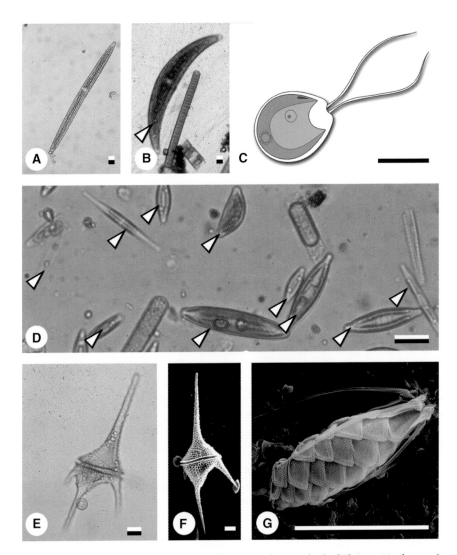

Figure 9.1 Unicellular algae consist of one cell and are microscopic. Scale bars = 10 microns. A and B: Species of *Closterium*, a common freshwater green alga, especially in nutrient-rich waters (light micrographs). C: *Chlamydomonas*, a common freshwater green alga that is motile by means of two flagella that pull it through the water (illustration). D: Various freshwater diatoms from a stream in Lancaster County, Pennsylvania, USA (light micrograph). E and F: Species of *Ceratium* from a freshwater pond in Pennsylvania (E) and Florida (F), USA (light micrograph, E, and scanning electron micrograph, F). G: A species of *Mallomonas*, a scaled synurid chrysophyte from freshwaters of Florida, USA (scanning electron micrograph). Micrographs in F and G provided courtesy of Akshinthala K. S. K. Prasad, Department of Biological Science, Florida State University. Photos in B, D, and E by James C. Parks.

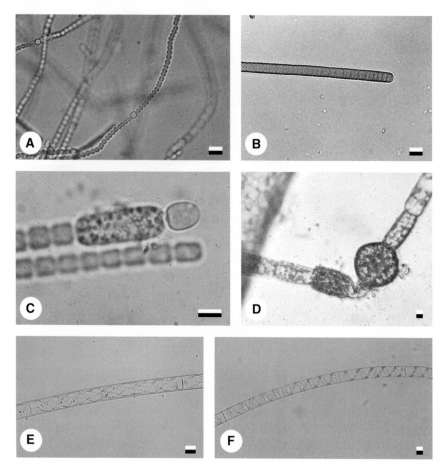

Figure 9.2   Filamentous algae are made up of a linear series of cells, some unbranched and others branched. Scale bars = 10 microns. A, B, and C: The freshwater blue-green algae (cyanobacteria) *Anabaena* (A), *Oscillatoria* (B), and *Cylindrospermum* (C). D, E, and F: The freshwater green algae *Oedogonium* (D) and *Spirogyra* (E and F). Photos in C and D by James C. Parks.

Filamentous algae consist of multiple cells but these cells are arranged only end-to-end such that their bodies consists of unbranched or branched filaments (Figure 9.2). Filamentous algae are abundant in aquatic systems and are commonly responsible for the green or blue-green mats seen floating at the surface of ponds or lakes. Others exist as short filaments that grow en masse as a "turf" on submerged objects such as rocks. Since algae come in several different colors, any green, red, or brown stains or patches evident on a crime suspect's clothing, for example, should be examined as potentially composed of filamentous algae picked up while in contact with water. Even if there are no visible patches of color, individual filaments are superficially similar to animal hairs in their fineness and the ease with which they can be picked yet go unnoticed by a criminal suspect, therefore forensic investigators should be prepared to have potential evidence examined microscopically for the presence of filamentous algae.

Larger, macroalgae come in a variety of more complex forms but all are readily visible to the naked eye as individual plants, unlike their filamentous or unicellular counterparts (Figure 9.3). Examples in freshwater include coleochaete and stoneworts. Examples in marine or estuary waters include sea-lettuce, rockweed, Irish-moss, sargassum, and the many different types of kelp. These large macroalgae are typically called "seaweeds" when they occur in marine waters.

Green, red, brown, golden-brown, and blue-green are the more common colors found in algae (Figure 9.3). Although all have the green pigment chlorophyll that makes photosynthesis possible, each group has a mix of accessory pigments that gives each its distinctive color. These distinctive accessory pigments are then used by algal taxonomists along with other cellular characteristics to recognize various formal taxonomic groupings of algae, some of which are described below.

Diatoms include at least 5000 species of algae that are mostly unicellular and golden-brown in color due to their brown accessory pigment, fucoxanthin, and possess a silica-based (glass-like) cell wall (Sze 1998; Figures 9.1D and 9.4). Some diatoms co-occur in colonies of cells (e.g., Figure 9.4F). An algal group closely related and biologically similar to diatoms, the chrysophytes, have silica-based scales around them instead of walls (e.g., Figure 9.1G; Sze 1998). The cell walls of diatoms (called frustrules) come in two halves that fit together like the two halves of a gift box (e.g., Figure 9.4A and C). The frustules themselves have an overall symmetry when viewed on their broad face (e.g., polysymmetric, Figure 9.4A–C and G; disymmetric, Figure 9.4D; monosymmetric, Figure 9.4E) and are highly ornamented with species-specific patterns of pores and grooves that can be discerned under compound light or scanning electron microscopy. Although some diatom species are terrestrial (e.g., in soils), the majority are aquatic (fresh- or saltwater) and grow on submerged surfaces, including those of other aquatic plants, in the intertidal sands of beaches, or as plankton suspended in the water column (Pollanen 1998; Sze 1998). Diatoms are the most commonly used algae in forensic investigations for the following reasons. First, their abundance in many different habitats means that they are usually present at a crime or accident scene and therefore are potentially available as trace evidence. Second, their hard, acid- and decay-resistant silica-based cell walls are reliably persistent and retain their species-specific diagnostic characteristics long after the actual cell has dried, died, or even been passed through the internal organs or digestive tracts of recovered bodies.

Green algae are the most abundant and diverse algae worldwide, with 7500 species. Their accessory pigments include the same mix of chlorophylls and carotenoids (e.g., β-carotene) found in land plants and therefore have the same "green" color that we have come to expect of plants. Although the carotenoids are generally yellow or orange, these algae appear green to us since the chlorophyll visually overwhelms or masks the carotenoids that also are present. Green algae are found in all types of waters and come in all forms, from unicellular (Figure 9.1A–C) and filamentous (Figure 9.2D–F), to the larger seaweeds (Figure 9.3G and H).

Red algae include 5000 species, most of which are marine seaweeds (Figure 9.3A–C). Their red color comes from their accessory pigments, called phycobilins, which mask the green of the chlorophyll also present. One common species is Irish-moss (*Chondrus crispus*), a red alga that grows in dense lawns attached

to rocks on rocky coasts in the intertidal zones along North America's and Europe's Atlantic coasts. A massive commercial industry has developed around the harvesting of this species (and some others) for the polysaccharides it produces, carrageenan and agar. Carrageenan is used as a thickener or stabilizing agent in dairy-based products such as ice cream and bottled chocolate milk. Agar is used as a thickener or solidifying agent in various food products such as vegetarian gelatin substitutes, ice cream, and jellies. It is also used in molecular biology to make the gels used in the analysis of DNA for forensic, diagnostic, or evolutionary studies.

Brown algae include 1500 species, most of which are marine seaweeds: the kelps, rockweeds, and sargassums are the most familiar (Figure 9.3A and D–F). Their brown color comes from the accessory pigment fucoxanthin. Many species, including the kelps, are edible and a large commercial industry has developed around the harvest of kelps (particularly off the west coast of North America) for the extraction of algin, a polysaccharide that has uses similar to carrageenan.

Dinoflagellates include over 3000 species, most of which (90%) exist as microscopic, unicellular, marine plankton. Some species are common in freshwater ponds, lakes, and bogs. Dinoflagellate cells have distinctive shapes and are typically armored with cellulosic plates called "thecae" (Figure 9.1E and F, Figure 9.5). The typical dinoflagellate is able to propel itself (e.g., towards light and nutrients) with its two flagella, one transverse flagellum that wraps around the cell equator like a belt and one longitudinal flagellum that hangs down from the cell. Dinoflagellates comprise an interesting group because only about half are photosynthetic phytoplankton, whereas the rest are colorless zooplankton that feed on other plankton. The photosynthetic dinoflagellates are among the most important primary producers in coastal waters. Dinoflagellates also form non-motile unicells called "cysts" as part of their lifecycle. The cyst of dinoflagellates is distinctive with its spiny wall protrusions and, internally, conspicuous red-pigmented structure. Cysts may remain dormant for many months or years and they accumulate in the bottom sediment of water bodies, making them common in water body sediment samples. Their abundance plus their resistance to damage by drying out and decay make them potentially of high value as trace evidence

---

Figure 9.3 (Please refer to the colour plate section.) Macroalgae are algae with bodies large enough to be seen to the naked eye. Most of these occur in marine waters and are called seaweeds. A: A mixture of mostly brown (rockweed, *Fucus*) and red (dulse, *Palmaria*) seaweeds in the intertidal zone of a rocky shore in Maine, USA. B: A mixture of mostly red and green seaweeds caught by trawling in the shallow waters of Wallops Island Inlet, Virginia, USA. C: A red seaweed, sea sacs (*Halosaccion*), in the intertidal zone along the coast in Vancouver Island, British Columbia, Canada. D: The brown seaweed sargassum (*Sargasso*) washed ashore on the Caribbean Sea in Aruba. E: The brown seaweed kelp (*Nereocystis*) washed ashore on the Pacific on Vancouver Island, British Columbia, Canada. F: Giant kelp (*Macrocystis*, a brown seaweed) growing beneath the surface of Uculet Inlet off Vancouver Island, British Columbia, Canada. G: Sea lettuce (*Ulva*, a green seaweed) caught by trawling in the shallow waters of Wallops Island Inlet, Virginia, USA. H: Chaetomorpha (*Chaetomorpha*, a green seaweed) in the intertidal zone along the Pacific coast of British Columbia, Canada. Photos in A, C, E, F, and H by James C. Parks.

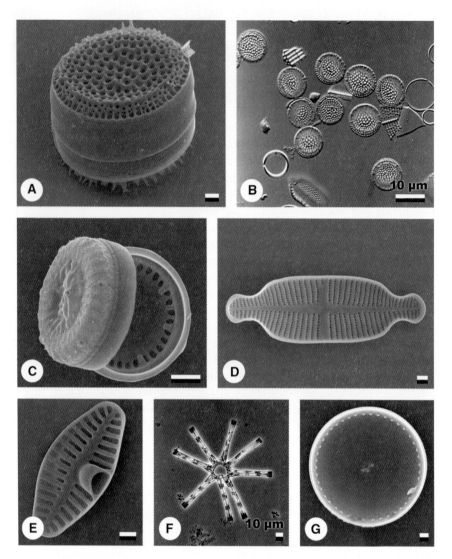

Figure 9.4 Diatoms come in a variety of forms and are found in a wide diversity of habitats. Scale bar = 1 μm, except where otherwise indicated. A–C: Some marine diatoms from coastal waters of south-eastern North America (A and B, *Thalassiosira cedarkeyensis*; C, *Cyclotella choctawhatcheeana*). D–G: Some freshwater diatoms from south-eastern North America (D, *Acanthidium exiguum*; E, *Planothidium lanceolatum*; F, *Asterionella formosa*; G, *Conticribra weissflogii*). A, C, D, E, and G are scanning electron micrographs. B and F are light micrographs using differential interference contrast and phase contrast, respectively. Photos courtesy of Akshinthala K. S. K. Prasad, Department of Biological Science, Florida State University.

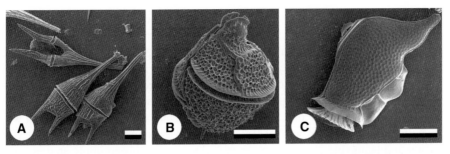

Figure 9.5 Dinoflagellates come in a variety of forms. Scale bar = 20 μm. *Ceratium hircus* (A), *Gonyaulax* sp. (B), and *Dinophysis caudata* var. *acutiformis* (C) from marine coastal waters of southeastern North America. Scanning electron micrographs courtesy of Akshinthala K. S. K. Prasad, Department of Biological Science, Florida State University.

in forensic investigations. Certain species are responsible for periodic blooms in coastal waters called "red tides" – the name alluding to the tinge of color they give to the water when in high density. Often red tides are accompanied by the beautiful phenomenon of bioluminescence, and occasionally they can be toxic and cause large fish die-offs, the closure of beaches and fishing areas, and contamination of shellfish stocks with potent neurotoxins capable of inflicting illness or even death in humans.

Blue-green algae (also known as blue-green bacteria or cyanobacteria) include at least 1500 species that are actually photosynthetic bacteria. They are distributed all over the world, throughout freshwater, estuarine, and marine environments. They come in various forms, from unicellular or filamentous forms (Figure 9.2A–C) to colonial forms

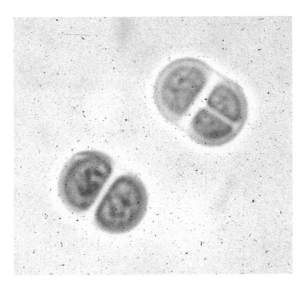

Figure 9.6 Colonies of the cyanobacterium *Anacystis* consist of two to several cells embedded in a gelatinous matrix and strongly contrast with the filamentous forms depicted in Figure 9.2A–C.

in which individual cells are arranged in sheets or small clusters (e.g., Figure 9.6), the individual cells being joined only by a gelatinous matrix they secrete. Many of the filamentous, mat-forming algae of pond surfaces, particularly if they are blue-green in color, are likely to be blue-green algae such as *Anabaena* (Figure 9.2A), *Lyngbya*, or *Oscillatoria* (Figure 9.2B). Many blue-green algae produce powerful dermatoxins, neurotoxins, or hepatotoxins – toxins that can make humans sick or prove fatal to livestock who have ingested or otherwise been exposed to water from a water body with a bloom (large growth) of toxic blue-greens (Hayman 1992; Codd *et al.* 1999; Falconer and Humpage 2006). Blue-green toxins have also been implicated in certain neurodegenerative diseases of humans, with some (e.g., lytico-bodig of Guam) occurring on the scale of regional or local epidemics that have lasted decades before the root cause of the disease has been identified (Cox and Sacks 2002).

Application of algal evidence in forensic investigations

Algae may potentially be of use to a forensic investigation in a variety of ways, four of which are discussed below: (1) distinctive algal species or communities of species carried on the person or property of a criminal suspect can link them to specific aquatic crime scenes or to physical evidence such as weapons discarded in water bodies, (2) the occurrence or lack of algae in bodily organs can be used for positive or negative diagnoses of death by drowning, and the relative abundances of diatoms in such tissues can be used to infer the probable location of drowning, (3) the discovery of toxic algae or their toxins in foods, drinking water, or swimming water can be used to determine the causal circumstances of human or domestic animal poisonings or deaths, and (4) ongoing studies of algal community succession or growth rates on submerged mammalian carcasses point to the possibility of using algae to estimate a minimum post-mortem submersion interval (PMSI).

(1) Linking suspects to specific aquatic crime scenes or physical evidence

Crime scenes in and around water are common. For example, it is not uncommon for the body of the victim of a homicide that occurred on land to be disposed of in a body of water. The goal of the killer may be to make the death *look* like a drowning or to hide the body. Similarly, murder weapons are often dumped by criminals into water bodies in order to prevent them from being found and used as evidence. Accidents in natural water bodies involving swimming, alcohol, and negligence or recklessness (possibly amounting to involuntary manslaughter) also occur. Other crimes that occur near water may be crimes of opportunity, such as a mugging where the assailant has found the victim walking on a deserted shoreline, out of sight or earshot of others.

In such cases, the concordance of specific types of algae (species) on the suspected criminal's clothing, vehicle, or person with the algal species growing at the scene of a crime or where physical evidence (such as a weapon) is dumped can provide valuable circumstantial evidence placing the suspect at the crime scene. Moreover, since various aquatic plants, algae, or their various reproductive parts are only present during specific times of year, the botanical evidence may be able to place the suspect at the given scene at the particular time of year the crime is said to have happened. In all cases, it is important to realize that algae found on a suspect do not have to be unique to the crime scene to place the suspect there; rather, a distinctive combination or one or two indicator algal species found on the suspect may provide strong circumstantial evidence that the suspect was at a particular pond or stream where the crime is said to have taken place. Alternatively, particular algal assemblages (i.e., groups of species) may not be specific to a specific locale, but rather to a particular *type* of water body (Siver *et al.* 1994; e.g., freshwater rather than marine, streams rather than ponds, or waters near sewage or thermal pollution sources, etc.).

Case study 1A

In the well-cited case of the *State of Connecticut (USA) vs. John C. Hoeplinger*, Hoeplinger was convicted of murdering his wife by blunt trauma to the skull on the morning of 7 May 1982 (Pagliaro 2005:179). The State used a filamentous alga found on Mr Hoeplinger's wet T-shirt to link him to the murder weapon, which had been hidden in a pond nearby. Prior to the discovery of the alga on the shirt, however, Mr Hoeplinger had denied the charges, saying instead that an unknown assailant must have killed his wife in or just outside their home. When police found the murder weapon (a brick with blood and hair from his wife embedded on it) in the nearby pond, Mr Hoeplinger denied having anything to do with it. When police inquired about why Mr Hoeplinger's shirt was wet, Mr Hoeplinger claimed to have rinsed it in tap water after accidentally getting his wife's blood on it on discovering her body. The Connecticut Forensic Science Laboratory used scanning electron microscopy to examine a visible deposit of green algae on the shirt: they identified on it the same filamentous green alga found growing in the pond where the murder weapon had been disposed of. Prosecutors and the jury concluded that Mr Hoeplinger had in fact rinsed his shirt in the pond water near to the time that he had disposed of the murder weapon (the brick) in the pond.

Case study 1B

In May 1992, three teenage boys were convicted of several felonies stemming from their involvement in the violent mugging of two younger boys who were fishing at a rural Connecticut, USA pond in July 1991 (Siver *et al.* 1994). In a presumed effort to steal the young boys' bicycles, the teenagers bound the victims with duct tape, beat them with baseball bats, and left them in the pond to drown. One of the victims managed to free himself, rescue his friend, and rally area neighbors for help. After a rapid investigation by police, the teenage assailants were apprehended and their wet sediment-encrusted shoes were seized for comparison to the algal assemblage found on the victims' shoes and in the pond at the crime scene. Using acid digests commonly used for diatom analysis, investigators at the limnological laboratory of Connecticut College identified the same peculiar ratio of dominant diatom and scaled chrysophyte species (diatom relatives) in all three samples. Based on this and other evidence, the court concluded that the teenage suspects were indeed guilty and they were convicted to serve a lengthy jail sentence.

(2) Diagnosis and placing of death by drowning

Diatoms are minute, ubiquitous algae present in all types of natural water bodies. When a person drowns, they invariably violently inhale water and, in natural bodies of water, this water will have diatoms in it (Figure 9.7). The force of this inhalation causes aveolar-capillary membranes in the lungs to rupture, allowing the diatom frustules to enter the bloodstream. The still-beating heart of the drowning victim then transports the diatoms in the bloodstream to organs in the body (Timperman 1972; Pollanen et al. 1997; Pollanen 1998; Keiper and Casamatta 2001; Aggrawal 2005). If the body had instead entered the water post mortem, water with diatoms may eventually have entered the lungs by sheer mechanical pressure alone, but those diatoms will not enter the bloodstream since the force of passive entry is too low for capillary rupture. Even if the diatoms were somehow to enter the capillary vessels of a dead body, the heart is not beating and the diatoms will not be transported to other tissues and organs (Figure 9.7). Although small numbers of diatoms are reportedly sometimes found in the organs of non-drowned victims, they are typically very low in abundance and are demonstrably from sources other than the water in which a body may have been dumped, for example diatomaceous earth (the fossilized remains of ancient diatoms) is used for its abrasive or sorptive properties in toothpaste, metal polish, facial scrubs, and household pest control products (Yoshimura et al. 1995; Pollanen 1998:6). The standard diatom test for confirmation of drowning is to sample marrow of an intact femur of the putative drowning victim and look for diatoms. Although other organs (e.g., kidneys, liver, or brain) may be sampled, femoral marrow is popular since the marrow of an intact femur is sealed from possible soiling from some external source of diatoms. If diatoms are found, then the case for drowning is strong and the investigation can then move to determining (i) where the drowning occurred, (ii) whether the drowning was accidental, suicidal, or homicidal, and (iii) how long the body had been lying in the water (Aggrawal 2005). A similar test for drowning using green algae, rather than diatoms, has been developed but it is not in widespread use (Yoshimura et al. 1995).

Pollanen (1998) discusses the importance of matching at least some of the species found in a decedent's femoral marrow to the drowning medium (i.e., the water in which the drowning occurred) in order to bolster the diagnosis of drowning. As Pollanen explains, the one caveat to this is that it will generally be only the smaller diatoms

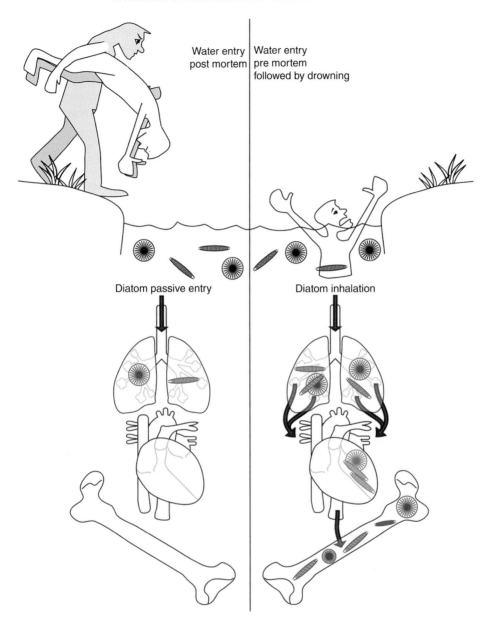

Water entry post mortem | Water entry pre mortem followed by drowning

Diatom passive entry | Diatom inhalation

Figure 9.7 The underlying principles of the diatom test for drowning.

(e.g., 10–50 μm diameter) that make it as far as the drowning victim's internal organs and so the diatom assemblage of the femoral marrow, for example, will typically be but a subset of that of the drowning medium. In addition to bolstering a diagnosis of drowning, the distinctive diatom assemblage found in internal organs may be used to infer where the drowning occurred, particularly in cases where there is good reason to believe the body has been moved post drowning to another location (Ludes *et al.* 1999). In general, the diatom assemblages of freshwater are substantially different to those of marine or estuarine waters, and differences in nutrient levels, temperature, and water flow rate among water bodies will conspire to produce different assemblages among even freshwater (or saltwater) water bodies (Siver *et al.* 1994). In an examination of 20

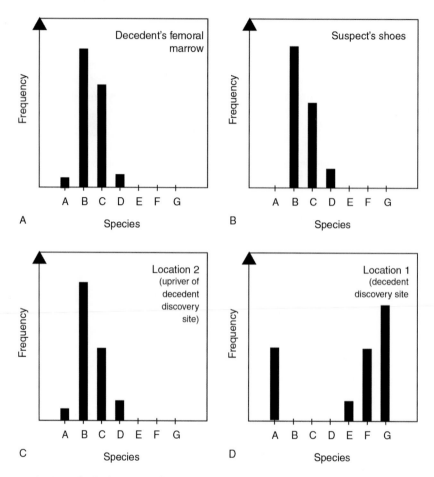

Figure 9.8 Hypothetical results of a diatom analysis in an investigation of a homicide by asphyxiation due to strangulation-facilitated drowning. In this example, the victim's body is discovered in a river at location 1 (D) with injuries to the neck indicative of strangulation. As a result of the action of water currents, the site where a victim's body is discovered is not always the site where the actual drowning occurred. In this hypothetical example, a match between the diatom assemblage of a drowning victim's femoral marrow (A) with the assemblage of location 2 (C) provides evidence that the drowning occurred at or near location 2, rather than where the body was discovered (location 1, D). Similarly, a match between the diatom assemblage found on the suspect (B) and the decedent's marrow (A) and location 2 (C) links the suspect to the crime.

corpses who died from accidental or suicidal drownings at known freshwater sites, diatoms were extracted from various organs of the deceased and the relative abundances of the major diatom species (i.e., the diatom assemblage) in the organ tissue were found to closely match the diatom assemblage of the water in which the drowning occurred (Ludes *et al.* 1999; see Figure 9.8). The authors concluded that distinctive diatom communities within victims' internal organs can assist forensic investigators in determining the location of drowning. Hürlimann *et al.* (2000) built on this by providing several useful ways of analysing such data statistically.

Case study 2A

In a case described by Aggrawal (2005), the diatom test was used to confirm death by drowning of a young woman from Hisar, India, whose body was found on 6 February 1992, some 350 km away in Lake Badkhal near Faridabad, India. She had last been seen three days earlier by her husband while they were both visiting relatives near Faridabad. The husband was immediately a suspect in her demise because he had been known to accuse her recently of infidelity and he had hit or otherwise been violent with her on several occasions. Furthermore, in a seeming display of guilt, the husband returned home from Faridabad without his wife yet did not bother reporting her as a missing person to the police. He told neighbors and police when they came asking several days later that he did not report her because he assumed that she had run off with her lover and because he no longer cared about his wife's whereabouts. Analysis of the corpse's stage of decay indicated that her body had only been submerged for about three days. The diatom test from femoral marrow provided a positive diagnosis of drowning. This is the extent of the algal evidence used in this case, but it is critical evidence since death by drowning makes it less likely that her husband killed her. This is because relatively few homicides of adults occur by drowning the victim because adult victims are usually assaulted some distance from a water body and they tend to resist being taken by their assailant to the water body unless they are first bound with ropes, knocked unconscious, or drugged. Analysis of the wife's body revealed no evidence of a struggle, being bound, or narcotics. Thus, investigators of this death ruled it most likely to be due to suicide and a case against her husband was not pursued.

Case study 2B

In a case described by Pollanen (1998:115) as "Murder in the Hudson River", analysis of diatom evidence at multiple levels was employed to convict a 63-year-old man from New York, USA, of murdering his 58-year-old live-in girlfriend. It was May 1996 when his girlfriend's recently dead body was recovered from the Hudson River of Ulster County, New York. The decedent had neck wounds consistent with

strangulation and toxicology of her blood revealed excessive levels of diphenhy-
dramine consistent with a drugging. Nitric acid extracts of the femoral bone marrow
revealed diatoms and comparison to the diatom community of the Hudson River
where she was found and that of her bathtub water revealed that she had drowned in
the river and not the bath, where drowning homicides often take place. The femoral
diatoms plus the strangulation wounds lead to a diagnosis of asphixia due to both
neck compression and drowning as the cause of death, with homicide the most likely
mode. Presented with this evidence, the man's nephew and apparent accomplice
testified against the man in exchange for immunity. The nephew indicated that the
suspect had drugged the woman and dumped her in the Hudson. Once in the river,
however, the drugged woman unexpectedly regained consciousness and the suspect
then waded out to strangle her as she drowned. Further testimony by the nephew lead
to the precise location along a road where the suspect had dumped his wallet, and the
watch and shoes worn during the murder. Distilled water washings of these articles
and subsequent nitric acid digestion of the washings revealed the same diatom
community as the presumed drowning site in the Hudson.

(3) Diagnosis of algaltoxin poisonings

Although most algae are not toxic, certain species of blue-green algae (cyanobacteria)
and dinoflagellates can produce toxins capable of causing illness or death in humans and
animals. Such poisonings are not common, particularly those of humans, and they will
typically fall outside of the realm of criminality; however, the forensic botanist, coroner,
and police investigator should be familiar with the circumstances under which they are
likely to occur.

   In freshwaters, the principal toxic algae are certain species of blue-green algae that
produce hepatotoxic, neurotoxic, or dermatoxic substances (generally called
"cyanotoxins;" Lee 1999; Graham et al. 2010). These cyanotoxins may reach harmful
concentrations during blooms (large, massive growths). These blooms are expansive
masses of algae near the surface of the water and their growth is fueled by the periodic
convergence of sufficient nutrients (e.g., fertilizer runoff from golf course lawns, soil
runoff from farms), warm temperatures, and sunlight. Farmers and veterinarians are
generally more familiar than most others with the danger posed by blooms since
livestock or other domestic animal illness and deaths that arise from drinking the
contaminated water bodies are common (Francis 1878; Stewart et al. 2006). Among
humans, exposure to cyanotoxins has come about by swimming in or drinking from
water with toxic blooms, or, in dialysis patients, dialysis for which the source water had
been contaminated. During investigations of mysterious illnesses, skin rashes, or the
rare death associated with nerve or liver damage, eyewitness accounts of "dirty" or
"scummy" green, brown, or blue-green masses in water from drinking reservoirs or
swimming holes to which the victim had been exposed should alert investigators to the
possibility that blue-green algae are the culprits. Because toxic blue-green algal blooms

also generally produce malodorous but non-toxic compounds, eyewitness accounts of the water having distinctive musty and/or earthy odors should also alert investigators to the possible involvement of blue-green algae (Graham *et al.* 2010). Although the precedent is to consider such poisonings as accidents, it is conceivable that property owners (including municipalities) could be held liable for poisonings stemming from afflicted water bodies that lie on their property.

In saltwaters, the principle toxin-producing algae are dinoflagellates. Although most dinoflagellates are harmless, the toxic species produce extremely potent neurotoxins called saxitoxins and gonyautoxins (Gessner *et al.* 1997; Food and Agriculture Organization of the United Nations 2004; Holstege *et al.* 2011). For convenience of discussion and in anticipation of future discoveries of a greater variety of dinoflagellate-borne toxins, the term "dinotoxin" is used here to refer collectively to toxins produced by dinoflagellates. It is during large blooms (i.e., red tides) of toxic dinoflagellates in coastal waters that sufficient amounts of dinotoxin can accumulate and lead to large fish die-offs and accumulation in shellfish such as oysters, clams, and mussels, due to the filter-feeding habit of shellfish. Therefore, during red tides, beaches are generally closed to swimming and afflicted waters are closed to shell-fishing since ingestion or consumption of dinotoxins by humans can lead to paralytic shellfish poisoning (PSP) and related conditions. Assuming oral ingestion of intox-icated shellfish, PSP symptoms start with numbness around the mouth and extending to the limbs, vertigo, incoherent speech, general nerve dysfunction, nausea, vomiting, and may lead to paralysis of the diaphragm and respiratory failure, leading to death within just hours of the earliest symptoms (Gessner *et al.* 1997; CDC 2011; Holestege *et al.* 2011). In early phases of PSP, the victim may appear to be in a drunken state. Although the US Food and Drug Administration guidelines call for the testing of all batches of shellfish sold within the United States, PSP in humans that have eaten intoxicated shellfish still occurs and could either be due to legally sold shellfish in which the toxins went undetected, illegally sold (i.e., not tested) intox-icated shellfish, or intoxicated private catches intended for private consumption. The US Centers for Disease Control and Prevention's Emergency Preparedness and Response guidelines list saxitoxin in particular as a hazardous, naturally occurring chemical capable of causing human health emergencies (CDC 2011). Furthermore, saxitoxin is listed as Schedule 1 under the Chemical Weapons Convention because of its potential use as a lethal chemical weapon (Holstege 2011). It is considered a potential bioterrorist threat in light of the relative ease of culturing dinoflagellates and the extreme potency of dinotoxins.

Case study 3A

Jochimsen *et al.* (1998) showed that water-borne cyanotoxins were responsible for the deaths of 26 of 130 patients that received hemodialysis over a four-day period from 17 to 20 February 1996 at a dialysis center in Caruaru, Brazil. The immediate cause of death was liver failure. Acute illness occurred in an additional 90 of the 130 patients, although these patients recovered. Enzyme-linked immunosorbent assay identified the presence of particular cyanotoxins called microcystins in water used for dialysis at the center, water in the source reservoir, and in the serum and liver tissue extracted from victims. Microcystins are powerful hepatotoxins produced by various species of the cyanobacterial genera *Microcystis*, *Anabaena*, *Oscillatoria*, and *Nostoc*, among others (Stewart *et al.* 2006). Pathological studies of the livers of 16 of the 26 victims showed extensive damage and cell death. This case highlights the need for extremely clean water for use during dialysis, since patients receive such massive amounts of water during this treatment.

Case study 3B

The *Milwaukee Journal Sentinel* newspaper reported in 2003 that the cause of the 2002 death of a Dane County, Wisconsin teenager was determined by the County Coroner to be heart failure caused by cyanotoxin poisoning (Behm 2003). In July of 2002, two boys became sick with diarrhea, acute abdominal pains, nausea, and vomiting. Within less than two days, one of the boys, 17-year-old Dane Rogers, went into shock and suffered a seizure before his heart failed. An autopsy by the University of Wisconsin–Madison Medical School offered no explanation, citing only acute heart damage and consequent cardiac arrest in a "...sad and vexing case." With laboratory tests for pesticides, parasites, and other disease organisms being negative, subsequent interviews with Dane's friends finally revealed that 48 hours before his death they had swum with Dane in a local golf-course pond to cool off amidst the summer heat. The boys elaborated that Dane and one other boy had been submerged for varying lengths of time and had also ingested some water from the pond. The boys' description of the pond at the time as "dirty" and "scummy" suggested to investigators that a toxic cyanobacterial bloom may be related to the death, since such blooms are frequent in golf course ponds and other water bodies with excessive amounts of fertilizer from runoff. Indeed, the freshwater cyanobacterium *Anabaena flos-aquae* and its toxin, anatoxin-a, were subsequently discovered in the blood and stool samples from both boys, and the toxin was at lethal levels in the decedent, Dane Rogers. Mysteriously, however, water samples collected and analysed from the pond just two weeks after the swim found neither cyanobacteria nor cyanotoxins, which could mean that the bloom had dissipated and the toxins degraded between the time of the swim and the pond water analysis. Nevertheless, after a year of puzzling investigation by multiple investigators, the County Coroner concluded that poisoning by anatoxin-a was responsible for death in the teenager.

Case study 3C

Gessner *et al.* (1997) provided a detailed clinical description of the onset and recovery of an accidental 1994 case of PSP of a man in Kodiak, Alaska. One or two hours after the consumption of beer and roasted mussels caught by himself from a beach on Kodiak Island, a healthy 28-year-old man was afflicted by acute tingling and numbness of the skin, nausea, and vomiting. Four hours later (five or six hours following mussel consumption) he was admitted to hospital as the symptoms progressed to severe headache accompanied by impaired speech, coordination, and swallowing. Within 15 minutes he experienced respiratory failure and had to be kept alive via a mechanical ventilator. Four hours after respiratory arrest, the patient was comatose and consideration was given to the possibility of brain death until a diagnosis of PSP advanced. With a diagnosis of PSP, intensive tests for brain death were deemed unnecessary and respiratory support was maintained with the understanding that he would fully recover relatively rapidly as the toxins were purged from his body. Gessner *et al.* (1997) emphasized the importance of the PSP diagnosis since, in areas where equipment and expertise for tests of brain death do not exist, a presumption of brain death may lead to wrongful (premature) termination of life support. Although the hospital's toxicology tests on the man did not test for dinotoxins, subsequent testing of mussels from the implicated beach revealed very high concentrations of dinotoxins. Gessner *et al.* stress the importance of healthcare professionals knowing the symptoms and appropriate management of PSP in order to avoid false diagnoses of coma or brain death, as well as to properly report and warn others in the area of dinotoxin emergencies in recreational waters.

(4) Estimation of post-mortem submersion interval

When a dead body is recovered and the circumstances of death or placement of the body are not known, it is useful not only to determine *how* the death occurred (i.e., the cause of death), but also *when* the death occurred (i.e., the post-mortem interval or PMI). If the decedent's identity is unknown, the PMI can guide the examination of missing person files in search for an identity. In homicide or manslaughter investigations, the PMI establishes a time frame within which a suspect would have to have been at the crime scene. Generally speaking, PMI can be measured in minutes, hours, years, or decades, depending on the physical condition of the body of the decedent.

There are a number of physical, bodily indicators of PMI such as livor (blood collecting/pooling towards gravity), rigor (body stiffening), and algor mortis (body cooling) that occur in the body within 24 hours after death. These phenomena are not reliable indicators of PMIs longer than 24 hours, however, and so other, longer-term indicators, such as the degree of decomposition by autolysis (breakdown of tissues by digestive fluids) or putrefaction (decay caused by the body's resident bacterial flora), are useful with certain important limitations whose discussion is beyond the scope of this chapter.

In terrestrial habitats, the existence of a carrion-feeding arthropod fauna (e.g., insects) has lead to the importance in recent decades of the forensic entomologist in helping to estimate a minimum PMI, or time frame from when the insects colonized the remains to the time of body discovery. In general, there is a well-understood chronological succession of arthropods that colonize to feed and/or lay eggs on a corpse following death: knowing the identity of the arthropods and their developmental stage provides the key to estimating a minimum PMI that can be measured on the order of minutes to months (Catts and Haskell 1990; Keiper *et al.* 1997).

In aquatic habitats, the goal is to estimate a postmortem submersion interval, or PMSI. The PMSI will be the same as the PMI when the death occurred in the water body and the corpse has been submersed in the water body up until its discovery by investigators. The PMSI will be different, however, when the death has happened some time before the submersion of the corpse in the water body. Nevertheless, the PMSI can be used in homicide investigations to establish a timeframe within which a suspect would have to have been at the water body to either drown or at least dispose of the body there. In the event that the decedent's identity is unknown and it is determined to be a death by drowning, the PMSI can guide the search of missing person records for an identity.

There are a number of physical, bodily indicators of PMSI that are useful for up to a week after submersion. In freshwater, wrinkling of the skin of the fingers, palms, and feet occurs within the first one to three days, and degloving (i.e., the skin of the hands and feet comes off like a "glove") after about a week of submersion (Pollanen 1998:30; Aggrawal 2005). Unfortunately, there appears to be no strong link between a carrion-feeding fauna and corpse decomposition in aquatic ecosystems. However, recent work has laid the foundation for efforts to develop the use of the algal flora that colonize the surface of submerged mammalian bodies to estimate a PMSI on the order of days, weeks, or more (Keiper and Casamatta 2001; Haefner *et al.* 2004; Zimmerman and Wallace 2008). The attributes of such algal communities (which include diatoms) that make them strong candidates for estimating a PMSI are as follows: (i) a ubiquitous distribution and year-round presence, (ii) the ability to grow on submerged surfaces such as corpses, (iii) their colonization of new surfaces follows patterns of ecological succession that are analogous to their terrestrial ecosystem counterparts, only in miniature, and (iv) their component species often have specific environmental tolerances, translating into distinctive algal species assemblages specific to particular types of water bodies. A study by Haefner *et al.* (2004), for example, found a strong correlation between chlorophyll-a concentration (an indicator of algal growth) on the surface of submerged piglet carcasses and PMSI in freshwater streams. A study by Zimmerman and Wallace (2008) found a strong correlation between diatom species diversity on submerged piglet carcasses and PMSI in estuarine waters. Outstanding areas of research lie with determining how robust the method of PMSI estimation from algal productivity or diversity statistics is to variation in parameters such as season and water body type or location. As this is an area of active research with many unanswered questions, these techniques have not yet been applied to casework.

Collection and processing of algal evidence in forensic investigations

The exact procedures for collecting algae will depend on which of the aforementioned applications are sought, the habitat sampled, and the type of algae sampled. Before collecting physical specimens, however, photographs, notes, drawings, and geocoordinates should be taken of the precise location from which the specimens are removed (e.g., a pond or a suspect's shoes) and of the physical appearance of the specimens at the time of collection since attributes such as color or cell shape may change dramatically before critical observation is eventually made by the forensic botanist. Assurance that samples are free of algal contamination is paramount since most algae employed in investigations are microscopic and therefore microscopic algal contaminants may go unnoticed and confound analysis of evidentiary samples (Table 9.1). As with any forensic investigation, it is vital to maintain appropriate chain of custody standards and

Table 9.1 Principles for the avoidance of contamination of evidentiary samples in forensic algal investigation

Principle	Methods to minimize violations of principle
1. Use reagents (e.g., distilled water, ethanol, soap used to clean instruments and glassware) that are free of microscopic algae.	a) Examine the source or stock of your reagents periodically with the compound microscope to ensure that they contain no algae.
2. Use clean and sterile equipment.	a) Clean and rinse all equipment with large volumes of algae-free (distilled) water.
	b) Consider using disposable pipettes.
	c) Communicate with maker of any disposable instruments for assurance that instruments should be free of algal contaminants.
	d) Periodically examine rinses of disposable equipment with compound microscope to ensure they are not contaminated.
3. Minimize the number of instruments (e.g., spatulas) and vessels with which your sample comes into contact during processing and examination.	a) Self-explanatory, since possibility of contamination increases with increasing number of instruments used.
4. Avoid cross-contamination of samples by not using the same sampling devices (e.g., pipettes, glass slides) for different samples.	a) Use disposable pipettes or spatulas.
	b) If not using disposable instruments, ensure that they are thoroughly cleaned before use with another sample.
	c) Change or dispose of instruments frequently.
5. Keep dated and chronological laboratory records of cases and results of tests.	a) Consecutive cases in which similar or identical algal species are discovered from evidence samples should lead to an investigation of possible laboratory sources of contamination.

records (Chapter 3) and it is crucial that all evidence be placed in clean containers and then sealed with the appropriate evidentiary tags (typically provided by the investigative police department), and labeled with the case number, victim's name, date, and location of sampling site.

Procedures for application 1: linking suspects to specific aquatic crime scenes or physical evidence

The forensic botanist should first survey the body of water or shoreline where the crime or accident is thought to have taken place and document the wetland and aquatic plant life there. Along the shore, trampled or recently disturbed patches of algae or other plants should be photographed, their location mapped, and then physically sampled as these areas may be the points of water body entry or egress by the perpetrator or victim. Entry and egress points of a water body will be rocky or sandy paths, soil, or through brush. The paths and soil may be lined with algae and other macroscopic wetland plants such as cattails, grasses, mosses, or liverworts found only along such waterways. The macroscopic plants can be collected and dried flat using a plant press (Chapter 3). Algae or very small aquatic plants growing on the surface of soil, sand, or rocks along the path, especially those associated with any footprints, may be sampled in a clean glass collection vial. The soil or sand itself may be sampled and processed for diatoms (Figure 9.9). Any rocks with noticeably disturbed patches of algae, mosses, or liverworts should be collected whole (if small enough) as these may represent areas where a suspect slipped or fell – a plausible happenstance given the slippery nature of such surfaces and the stress the suspect likely experienced while conducting the illegal activity. The collection of such rocks should be coordinated with human forensic investigators involved, and precautions to avoid contamination with human DNA

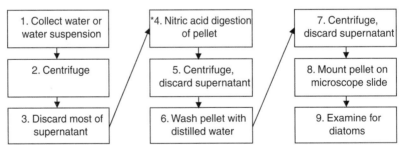

*To be conducted in fume hood.

Figure 9.9 Nine-step overview of acid digestion to extract diatoms from water or muddy suspensions for various subsequent applications (see Pollanen (1998:90) for details). Water or watery suspension may be water straight from the water body or from a distilled water wash of putative physical evidence. Washing of an object should be by shaking in a clean, watertight, plastic bag with just enough water to cover the object.

should be taken since the suspect may have left some DNA through skin or blood on these rocks during the fall.

All of these samples, of course, are collected for comparison and possible matches to any algal or plant fragments found on areas of the suspect or physical evidence. On the suspect, areas on which to focus sampling include the shoes (including but not limited to the outsole), socks, pant bottoms/cuffs, or higher up on the pants if a fall by the assailant is suspected. For a vehicle suspected to have been used in the crime, the driver-side floor, floor-mat, and foot pedals are places where algae/plants are likely to have been rubbed off shoes. Floors or trunk bottoms elsewhere in the car that carried tools or garbage related to the criminal act may also hold algae/plant materials. Tire treads and vehicle undercarriages or wheel wells are places where the vehicle may pick up and carry away the very same algae/plants, often encrusted in dried mud. Macroscopic algae or plant fragments on any of these items can be removed from the item whole using forceps and placed into clean glass vials for later identification. Smaller, typically microscopic remains of algae can be extracted from bulky samples of dried mud or sand removed from these items by resuspending them in clean glass vials using distilled water or 70% ethanol for subsequent microscopic examination. Physical evidence can also be rinsed with distilled water or 70% ethanol to collect any microscopic algal or plant material that may be attached. If the physical evidence is small enough, it can be placed in a clean plastic bag containing just enough distilled water or ethanol to cover it, and then shaken to extract any microscopic algal or plant material.

From the water, samples should be taken from the water surface, water column, and possibly bottom surfaces at multiple locations in a pond, for example, or further upstream in riverine systems since dead bodies or physical evidence can float or be carried distances from the original entry into the water and the botanist will want to know for later analysis how algal communities change in species composition and relative abundances from one site to the next. This should be done as close to the actual time that the crime is said to have taken place, but this is not absolutely necessary since a knowledgeable botanist can usually reconstruct the aquatic flora of the crime scene at the time of the crime based on a later survey.

Macroscopic algae or masses of algae may be visible at or near the water surface in the form of filamentous pond-scum, mat-forming algae, or floating stoneworts or seaweeds that have become detached from their submerged substrates. Macroscopic algae may also be found growing attached to a submerged substrate and this attached algae is of particular interest if found in the vicinity of putative entry and egress points to the water body. If possible, this material should be collected and preserved via two methods: (i) pressing and drying as herbarium specimens and (ii) wet samples in clean glass vials, with subsequent liquid preservation if deemed necessary. To make a herbarium specimen, a mass of filaments or macroscopic stonewort or seaweed is floated onto a sheet of paper (typically herbarium-grade, cotton-rag, heavyweight paper) by lifting the paper from beneath the specimen. To make this easier, the portion of the algae to be collected may be removed from the water first and placed into a 12 × 18" pan of water with a wetted herbarium paper sheet at the bottom. The paper is slowly

Table 9.2 Some commonly used preservatives for the preservation of algae. All of these preservatives will prevent the decay and decomposition of the alga in question, yet various artifacts will be had, depending upon the preservative used

Preservative	Notes	Recipe
Formaldehyde solution (Britton and Greeson 1987)	Preserves color but may distort cell shape and cause loss of any flagella that are present.	Per 100 ml of sample, add 3 ml of formalin, 0.5 ml of 20% detergent solution, and 0.1 ml cupric sulfate solution.
Lugol's solution (Britton and Greeson 1987)	Preserves cell morphology and flagella but stains cells brownish yellow.	Per 100 ml of sample, add 1 ml of Lugol's solution (120 ml of Lugol's solution is made by dissolving 5 g of I_2 (pure iodine) and 10 g of KI (potassium iodide) in 100 ml of distilled water, then mixing in 10 ml of concentrated glacial acetic acid).

raised and the alga comes to rest on its surface. The alga will naturally stick to the paper, after which it can be left to dry and it will be preserved indefinitely for subsequent examination and vouchering. Wet-grab samples should be examined for the identity of algal species by the forensic botanist as soon as possible. If the alga cannot be examined within 24–48 hours, or if it is necessary to use the wet grab sample as a voucher, long-term preservation can be achieved by using a preservative (Table 9.2).

Samples of free-floating, microscopic algae (e.g., unicellular algae such as diatoms, dinoflagellates, or fragments of filamentous forms) may be obtained via grab samples collected from glass jars from and below the water surface. The microscopic algae in these samples will be various and the water samples may be examined microscopically for these various species as is or following preservation (Table 9.2). However, a forensic botanist may choose to investigate only the diatom flora of the water because diatoms are abundantly present in aquatic ecosystems and, whereas most other algae will be difficult to identify in their dry state when collected from suspects or physical evidence, the species-specific diagnostic attributes of diatoms are found in their frustules, which are resilient to drying or other abuses and therefore persistent indefinitely. If diatom evidence is sought, somewhat large volumes (500–1000 ml) of water should be collected in clean glass jars. Although lower volumes of water may suffice, the higher volumes are preferred since diatom abundance can be low depending on the season, time of day, or temperature (Pollanen 1998). After collection, processing for examination typically involves concentration of the diatoms via centrifugation, a nitric acid digest of the pellet (the "pellet" being the settled material – including the diatoms – following centrifugation), followed by repeated distilled water washings and centrifuge-mediated concentration of the diatoms prior to mounting of the diatoms on a microscope slide for examination under compound light or scanning electron microscopy (Figure 9.9). The nitric acid digestion does not dissolve or degrade the diatom frustule, yet it does make the diagnostic features of the frustules easier to see and it dissolves other, non-diatom organisms.

Samples of microscopic unicellular or turf algae attached to bottom substrates may be desired, particularly where they occur at shallow enough depths that they may be relevant to the case (e.g., where a suspect might have walked and disturbed the substrate). In such cases, samples from multiple locales may be obtained by wet-grab samples from a loose, sedimentary substrate using a clean glass jar. If the algae are tightly attached to hard surfaces (e.g., rocks or wood), a specially designed sampling syringe can be used (see Zimmerman and Wallace 2008, Haefner *et al.* 2004).

These samples from the water body are then examined for possible matches to any algal or plant fragments found on the suspect or physical evidence. The same procedures are used to sample from the suspect or physical evidence as discussed earlier for shoreline botanical evidence. If diatoms were the focus of examination of the water samples, then diatoms will also be the focus during examination of the suspect or physical evidence, and the same general procedures involving nitric acid (or sulphuric acid) digestion may be followed (Figure 9.9). Rather than being washed, samples of textiles may be dissolved directly in nitric or sulphuric acid (Uitdehaag *et al.* 2010) and the procedure run as in Figure 9.9.

Procedures for application 2: diatom test for drowning

In order to robustly apply the diatom test for drowning, samples should be taken of the putative drowning medium at the scene of body recovery and the marrow of an intact femur from the body at autopsy (Pollanen 1998:83).

Samples of the putative drowning medium (i.e., the water) should be taken from the scene of body discovery at the time it is discovered or as soon thereafter as possible. Water samples should be placed in clean glass jars. Since planktonic diatom abundance can be low depending on the season, time of day, or temperature, it is recommended that 500–1000 ml of water be taken in order to have sufficient diatoms for examination after processing (Pollanen 1998). If planktonic diatom abundance is high, lower volumes of samples can be justified. If the body is recovered in very deep water, one should employ the services of a police diving unit to obtain samples at multiple depths in the water column. If there is suspicion that the exact location of body recovery is not where the body entered the water and may have drowned (e.g., river currents have moved the body), then an attempt to locate a probable area of entry (e.g., as indicated by footprints or disturbed path along the shore) should be made and water sampled from that area too. After collection, processing and examination of the water for diatoms should proceed according to standard protocols.

If samples of the putative drowning medium are unavailable or the drowning site unknown, it is possible that samples of the drowning medium can be obtained from bodily fluids (Pollanen 1998:87). The stomach, for example, may have substantial amounts of water that was swallowed by the victim during drowning. Although the finding of diatoms in such bodily fluids is not sufficient to diagnose drowning (e.g., water may enter the stomach passively following death), a match between diatoms found in the femur and stomach, for example, will bolster the diagnosis of death by drowning.

An intact femur of an adult may contain about 50 g of marrow and all of it should be removed after cutting of the femur at post mortem examination. This marrow is then digested with a strong acid (e.g., concentrated nitric acid), the acid digest centrifuged, most of the supernatant removed, and then the precipitate repeatedly washed and centrifuged in preparation for observation with the compound or scanning electron microscope and search for diatoms (Figure 9.10).

Procedures for application 3: diagnosis of algaltoxin poisoning

Aside from the recognition of symptoms discussed earlier, evidence for the confirmation and location of algaltoxin poisoning requires that samples be taken from (i) any visible growths of algae at or near the surface of the supposed toxic water body, (ii) the water of the suspected water body, and (iii) the urine, blood serum, tissues, or stool of the poisoning victim. In the case of dinotoxin poisoning from shellfish, a test for the presence of dinotoxins in intoxicated shellfish is also important. Water capable of causing acute sickness or the rare death due to algaltoxins will have noticeable blooms of the culprit alga present near the surface during the day when oxygen production via photosynthesis is sufficient to make the algal mass buoyant. Such blooms and the toxins they produce can be short-lived, particularly for freshwater blue-green algal blooms, and so it is important to collect water from the supposed toxic water body as soon as possible after exposure of the victim to the water.

In freshwater, only certain blue-green algae and the cyanotoxins they produce present health problems for humans and domestic animals (Graham *et al.* 2010). During investigations of mysterious illnesses, skin rashes, or the rare death associated with nerve or liver damage, eyewitness accounts of "dirty" or "scummy" green, brown, or blue-green masses in water from drinking reservoirs or swimming holes should alert

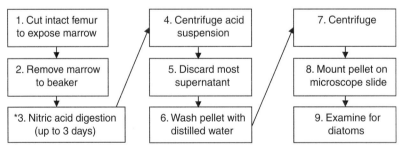

*To be conducted in fume hood.

Figure 9.10 Nine-step overview of acid digestion to extract diatoms from femoral bone marrow for the subsequent application of the diatom test for drowning (see Pollanen (1998:89) for details). Note that steps 3–9 in this protocol match steps 4–9 in Figure 9.9.

investigators to the possibility that blue-green algae are the culprits. Because toxic blue-green algal blooms also generally produce malodorous but non-toxic compounds, eyewitness accounts of the water having distinctive musty and/or earthy odors should also alert investigators to the possible involvement of blue-green algae (Graham *et al.* 2010). When sampling from algal mats and water suspected of containing cyanotoxins, one should take precautions (such as wearing latex gloves) to avoid contact between the water and one's skin, mucous linings of the nose or eyes, or ingestion orally. Single wet-grab samples in clean amber glass bottles each of 500–1000 ml from multiple locations in the suspected water body – particularly areas where there are visible concentrations of algae – are desired. The amber glass prohibits the light-induced degradation of cyanotoxins. Whereas a forensic botanist will generally be capable of identifying the species of algae in the sample, they will generally not have the ability to conducting the biochemical assays for the detection of various cyanotoxins. Thus, the forensic botanist or investigative agency should have the name and address of nearby laboratories capable of conducting such assays. The names of such laboratories may be obtained from most municipal public works departments since, while they do not generally conduct the tests themselves, they know who to send water samples to in the event that they do need tests conducted on public water supplies. Should the reader be interested in learning these techniques themselves, Graham *et al.* (2010) may be consulted.

In saltwater, the principle toxin-producing algae are dinoflagellates and the toxins generally only become a health concern when they reach harmful concentrations, such as during a red tide. It is important to note, however, that not all red tides are toxic and not all toxic dinoflagellate blooms need be red (Ely and Ross 2009). Thus, while description of the color of the water body can be useful in diagnosing a bloom, the color in and of itself is not sufficient grounds for confirming its toxicity. Poisoning could conceivably occur after exposure during recreational swimming in a toxic algal bloom, but more commonly occurs after eating contaminated shellfish since shellfish accumulate toxins during toxic blooms. Sampling protocols are the same as above for blue-green algae and cyanotoxins, although dinotoxins are not light sensitive and so amber glass bottles need not be used (Food and Agriculture Organization of the United Nations 2004). Although the dinoflagellate species may generally be identifiable by a capable forensic botanist, testing for the presence of dinotoxins must usually be conducted by a capable laboratory specializing in environmental toxicology.

Procedures for application 4: PMSI estimation

Sampling for PMSI estimation varies depending on the approach used. Generally, however, the approach taken is to ensure that the algal sample from the surface of a corpse is not contaminated from algae elsewhere (such as the water column or another submersed surface). Haefner *et al.* (2004) and Zimmerman and Wallace (2008) used a specially adapted syringe that creates a seal over the surface of the carcass while an internal brush sweeps the surface clean for the removal of algae attached to the carcass surface. Afterwards, Haefner *et al.* used standard protocols for quantifying chlorophyll-a concentration (an indirect measure of the amount of algae on the carcass), whereas Zimmerman and Wallace used standard acid digests and compound microscopy to assess diatom species diversity and ecological succession on the piglet carcasses used in their study.

Acknowledgements

The authors thank Akshinthala K. S. K. Prasad for many of the high-quality micrographs of algae used here, as well as Austin Mast for helpful discussion during the preparation of this chapter.

References

Aggrawal, A. (2005) Death by drowning. *Web Mystery Magazine*, 3, 1. Available at: http://lifeloom. com/III1Aggrawal.htm. Accessed 20 November 2010.

American Water Works Association (2002) Identification of algae in water supplies (CD-ROM). Available at: http://www.awwa.org/.

Behm, D. (2003) Coroner cites algae in teen's death: experts are uncertain about toxin's role. *Milwaukee Journal Sentinel*, 6 September.

Britton, L.J. and Greeson, P.E. (1987) Methods for collection and analysis of aquatic biological and microbiological samples: Phytoplankton. In: Britton, L.J. and Greeson, P.E. (eds), *US Geological Survey, Techniques of Water Resources Investigations*, Book 5, Chapter A4. USGS, Washington, DC, pp. 99–116.

Catts, E.P. and Haskell, N.H. (1990) Entomology and death: a procedural guide. Joyce's Print Shop, Inc., Clemson, SC.

CDC (2011) Saxitoxin. In: Centers for Disease Control & Prevention Emergency Preparedness & Prevention Guidelines. Available at: http://www.bt.cdc.gov/chemical/. Accessed 30 December 2010.

Codd, G.A., Bell, S.G., Kaya, K., Ward, C.J., Beattie, K.A., and Metcalf, J.S. (1999) Cyanobacterial toxins, exposure routes and human health. *European Journal of Phycology*, 34, 405–415.

Cox, E.J. (1996) *Identification of freshwater diatoms from live material*. Chapman & Hall, London.

Cox, P.A. and Sacks, O.W. (2002) Cycad neurotoxins, consumption of flying foxes, and ALS-PDC disease in Guam. *Neurology*, 58, 956–959.

Dillard, G.E. (1999) *Common freshwater algae of the United States, an illustrated key to the genera (excluding diatoms)*. J. Cramer, Stuttgart.

Ely, E. and Ross, N.W. (2009) Red tide in the Northeast. Rhode Island Sea Grant Factsheet P1099. Available at: http://seagrant.gso.uri.edu/factsheets/index.html.

Falconer, I.R. and Humpage, A.R. (2006) Cyanobacterial (blue-green algal) toxins in water supplies: Cylindrospermopsins. *Environmental Toxicology*, 21(4), 299–304.

Food and Agriculture Organization of the United Nations (2004) Marine biotoxins. *FAO Food and Nutrition Paper* 80.

Francis, G. (1878) Poisonous Australian lake. *Nature*, 18, 11–12.

Gessner, B.D., Middaugh, J.P., and Doucette, G.J. (1997) Paralytic shellfish poisoning in Kodiak, Alaska. *Western Journal of Medicine*, 166, 351–353.

Graham, L.E. and Wilcox, L.W. (2000) *Algae*. Prentice Hall, Upper Saddle River, NJ.

Graham, J.L., Loftin, K.A., Meyer, M.T., and Ziegler, A.C. (2010) Cyanotoxin mixtures and taste-and-odor compounds in cyanobacterial blooms from the midwestern United States. *Environmental Science Technology*, 44, 7361–7368.

Haefner, J., Wallace, J.R., and Merritt, R.W. (2004) Pig decomposition in lotic aquatic systems: the potential use of algal growth in establishing a postmortem submersion interval (PMSI). *Journal of Forensic Science*, 49(2), 330–336.

Hayman, J. (1992) Beyond the Barcoo – probable human tropical cyanobacterial poisoning in outback Australia. *Medical Journal of Australia*, 157(11–12), 794–796.

Holstege, C.P. (2011) Saxatoxin. In: Holstege, C.P., Neer, T.M., Saathoff, G.B.,and Furbee, R.B. (eds), *Criminal Poisoning: Clinical and Forensic Perspectives*. Jones and Bartlett, Sudbury, MA.

Hürlimann, J., Feer, P., Elber, F., Niederberger, K., Dirnhofer, R., and Wyler, D. (2000) Diatom detection in the diagnosis of death by drowning. *International Journal of Legal Medicine*, 114, 6–14.

Jochimsen, E.M., Carmichael, W.W., An, J., Cardo, D.M., Cookson, S.T., Holmes, C.E.M., de C. Antunes, M.B., de Melo Filho, D.A., Lyra, T.M., Barreto, V.S.T., Azevedo, S.M.F.O., and Jarvis, W.R. (1998) Liver failure and death after exposure to microcystins at a hemodialysis center in Brazil. *New England Journal of Medicine*, 338, 873–878.

Keiper, J.B. and Casamatta, D.A. (2001) Benthic organisms as forensic indicators. *Journal of the North American Benthological Society*, 20, 311–324.

Keiper, J.B., Chapman, E.G., and Foote, B.A. (1997) Midge larvae (Diptera: Chironomidae) as indicators of postmortem submersion interval of carcasses in a woodland stream: a preliminary report. *Journal of Forensic Sciences*, 42, 1074–1079.

Lee, R.E. (1999) *Phycology*, 3rd edn. Cambridge University Press, Cambridge.

Ludes, B., Coste, M., North, N., Doray, S., Tracqui, A., and Kintz, P. (1999) Diatom analysis in victim's tissues as an indicator of the site of drowning. *International Journal of Legal Medicine*, 112, 163–166.

Pagliaro, E. (2005) Additional case studies, Chapter 11, part 2. In: H.M. Coyle (ed.), *Forensic Botany: Principles and Applications to Criminal Casework*. CRC Press, Boca Raton, FL, pp. 179–183.

Palmer, C.M. (1959) Algae in water supplies: An illustrated manual on the identification, significance, and control of algae in water supplies. *Public Health Service Publication* no. 657 Washington DC. US Department of Health, Education, and Welfare Public Health Service.

Patrick, R. and Reimer, C. (1966) *The Diatoms of the United States*. Livingston Publishing Company, Philadelphia.

Pollanen, M.S. (1998) *Forensic Diatomology and Drowning*. Elsevier, Amsterdam.

Pollanen, M.S., Cheung, L., and Chaisson, D.A. (1997) The diagnostic value of the diatom test for drowning. I. Utility: a retrospective analysis of 771 cases of drowning in Ontario, Canada. *Journal of Forensic Sciences*, 42, 281–285.

Siver, P.A., Lord, W.D., and McCarthy, D.J. (1994) Forensic limnology: the use of freshwater algal community ecology to link suspects to an aquatic crime scene in southern New England. *Journal of Forensic Science*, 39, 847–853.

Stewart, I., Webb, P.M., Schluter, P.J., and Shaw, G.R. (2006) Recreational and occupational field exposure to freshwater cyanobacteria – a review of anecdotal and case reports, epidemiological studies and the challenges for epidemiological assessment. *Environmental Health: A Global Access Science Source*, 5, 6.

Sze, P. (1998) *A Biology of the Algae*, 3rd edn. McGraw-Hill, Boston.

Timperman, J. (1972) The diagnosis of drowning. A review. *Forensic Science*, 1, 397–409.

Uitdehaag, S., Dragutinovic, A., and Kuiper, I. (2010) Extraction of diatoms from (cotton) clothing for forensic comparisons. *Forensic Science International*, 200, 112–116.

Yoshimura, S., Yoshida, M., Okii, Y., Tokiyasu, T., Watabiki, T., and Akane, A. (1995) Detection of green algae (Chlorophyceae) for the diagnosis of drowning. *International Journal of Legal Medicine*, 108, 39–42.

Zimmerman, K. and Wallace, J.R. (2008) Estimating a postmortem submersion interval using algal diversity on mammalian carcasses in brackish marshes. *Journal of Forensic Science*, 53, 935–941.

10 Case studies in forensic botany

David W. Hall, Ph.D.

Placing people or objects at scenes

Case study 1

A woman was kidnapped, taken to a secluded wooded area, and sexually assaulted. With the assistance of a law enforcement officer familiar with the area, she was able to find the scene of the attack. The victim was also able to find the exact spot of the assault. It was outside a city in a normally wet area that was dry at this time. The examination of the spot revealed thousands of seeds and broken flowering parts. These seeds and flower parts were sampled and collected. The surrounding vegetation was also sampled so that the scene could be documented and separated from similar areas.

On arrest of a suspect, a blanket described by the victim as being used in the attack was found. The blanket was covered with plant material (Figure 10.1). On examination, thousands of seeds and flower parts were found. Seven different species of plants were represented. Four of the species were widespread weeds found in most lawns and open disturbed areas. These four species were represented by only a few samples among the thousands on the blanket. The remaining thousands of seeds, flower parts, and fragments were the same species as those found at the scene of the attack, which had previously been examined.

The suspect said he had used the blanket for picnics with his family on his lawn and at two local lakes. These areas were checked for evidence of the three plants. One of the species was found as a weed in the lawns. It has light, hairy, wind-blown seeds, and frequently occurs in lawns, but is killed by mowing long before it would mature and produce flowers and fruits.

Forensic Botany: A Practical Guide, First Edition. David W. Hall and Jason H. Byrd.
© 2012 John Wiley & Sons, Ltd. Published 2012 by John Wiley & Sons, Ltd.

Figure 10.1 A blanket with many adhering plant fragments was utilized in a sexual assault case to link the suspect to the scene.

At the trial the three kinds of plants on the blanket were described as matching those at the scene. It was also explained that the plants grew in moist to wet areas and would not be found growing to maturity in lawns. The defense maintained that the seeds and other fragments could have blown into those lawn areas near lakes where the blanket was said to have been used. It was pointed out that two of the species did not have seeds which could be carried by the wind. There were three clinching facts: (i) the plants have to be mature and are three to nine feet tall when flowering and fruiting, which would not happen in a mowed lawn, (ii) the thousands of seeds and fragments would not be found in a lawn where the mature plants did not grow, and (iii) these seeds and fragments matched those at the scene of the attack. The defendant was convicted.

Case study 2

In April 1998 a woman's body was found in a ditch on the south side of Spokane, Washington, about 50 feet from where two other women's bodies had been found four months earlier. All three bodies had been buried in plant material, which was collected and stored in the sheriff department's evidence room. In May, Dr Richard Old was called in to examine the plant material to see if it would provide any meaningful information.

The material consisted of the deciduous parts of horticultural plant material as well as a large volume of chipped "landscaping bark" (Figure 10.2). While the material from the two dump sites displayed a very different species composition in the majority of the material, there were also minor amounts of cross-contamination throughout the samples that did not occur at the site where the bodies had been dumped. The samples also failed to yield any evidence of grass clippings or needles from ponderosa pine, the most common landscaping tree in Spokane.

Based on the material, stage of decomposition, and degree of cross-contamination, it was determined that the material was likely raked from two separate "bark beds" in the same yard during the fall of 1997. The dominant and secondary species in each bed were

Figure 10.2 Material buried with a body linked the remains to one of two "bark beds" located within the same yard.

identified (a total of 14 species). The presence of particular species their spatial correlation, the bark material, and the absence of some species were all considered, the results produced a picture of a yard that would be very uncommon if not unique.

Several different ways of using this evidence were considered, including infrared aerial photography to identify the particular combination of tree species, and having a survey of neighborhood yards conducted under the guise of collecting data on what were the most common landscaping plants in the city. Both of these techniques were quite expensive and were not utilized, since there was the possibility that the material might not be from the killer's yard as it could have come from a landscaping service or been picked up in bags left out for disposal.

Instead, pressed specimens of each species were provided to the detectives who were interviewing suspects at their homes, and shown to potential witnesses. Two years later, when the correct suspect was arrested, analysis of his yard showed that it was indeed the source of the plant material used to conceal the bodies.

Case study 3

A lady visiting in an apartment complex walked across the complex to do laundry. Returning to her apartment she had a brief glimpse of a man who had a pillowcase over his head. She immediately escaped into the common area of the complex. A male suspect, having watched the woman escaping, was seen running out of the apartment and leaving. The man was observed to have picked up a motorcycle helmet from beneath a bush. A motorcyclist was stopped nearby minutes later; he had a leaf fragment beneath his visor.

The leaf fragment had enough characteristics present to enable it to be identified. It was the same kind of plant as that growing where the helmet was picked up. Unfortunately, the fragment did not show any cut or broken surface in common with the plant at the scene. Some intermediate piece was probably lost along the way. Unfortunately for the suspect the plant at the scene was unusual in that it is seldom grown as an outdoor plant in that area. The area is subject to freezing temperatures and this plant easily freezes. It could only grow outdoors in a spot protected from freezing. This plant was in such a protected spot and its size indicated that it had grown outdoors for several years.

This unusual situation provided strong circumstantial evidence linking the suspect to the scene. The suspect was convicted and sentenced.

Case study 4

A murder victim was found in his barn. He had been killed with a shotgun. A young next door neighbor was immediately a suspect because of a recent heated argument. A shotgun in the suspect's house had been recently fired. Also, muddy footprints led from the barn to the suspect's house. A muddy pair of the suspect's shoes was found in the house. On arrest the suspect confessed, but recanted when advised by a lawyer. The district attorney was now faced with the need for additional evidence.

The butt of the gun had several small plant fragments attached. All the plant fragments were determined to be one kind of grass. Grasses are particularly difficult to determine from fragments. Luckily one fragment was of a portion of a leaf with a distinguishing hair characteristic. The grass was a common species used for forage, hay, or lawns, and also found in most disturbed habitats (Figure 10.3).

Figure 10.3 Fragment of dried Bermuda grass lodged in the butt plate on the stock of a shotgun was linked to a hay barn.

The fragments, when found shortly after the shooting, were already dried, so had not been stuck to the gun while green, i.e., pulled from a growing plant. In the several county regions around that of the murder this grass is seldom used for forage, hay, or lawns. The murdered farmer had used the grass for hay and had placed a pile of it outside the barn window through which he was shot. The killer had apparently put the gun down in the hay while he went into the barn to see if the victim was dead. This strong circumstantial evidence helped lead to a conviction and sentence.

Case study 5

In northern Idaho, a man contacted authorities saying that his wife had gone huckleberry picking in the mountains the day before and had not returned. After two days of searching the area where she often picked huckleberries, her body was found at the base of a steep slope below a logging road. Her vehicle was missing, and it was assumed that it had been taken by the killer, and that possibly the theft was the motive for the murder.

The body had rolled down the bank, accumulating a considerable amount of plant material. All of the plant material was collected from the body and sent to investigators for analysis in hopes that any plant material from a different site could be found to check if she had been killed somewhere else and the body dumped at this spot. Plant fragments from her hair and inside the back of her blouse revealed both lawn clippings and petunia leaves and flowers (Figure 10.4a,b).

The back sidewalk of her residence was found to be lined with petunias and further investigation showed the remnants of a large amount of blood which had been poorly cleaned from the living room carpet. After questioning several residents of the small town, the wife's car was found in a rented storage facility, a few blocks from her home. The husband was subsequently charged with her murder.

Case study 6

An intact skeleton was found at an airplane crash site. The circumstances were such that the remains could not have been part of the crash. The crash was so severe that anyone in plane or thrown from it would have many broken bones. Also, the plane was full of fuel and burned fiercely. The pilot's remains found in the plane were severely burned. Some remains were only ash. The skeleton found outside the plane was completely clean with no burns and no broken bones. A small patch of skin behind an ear against the ground had human hairs and a few plant fragments stuck to it. The plant fragments luckily included one portion of a grass seedhead from which a definitive identification could be made. This grass is used for forage and is not very common. All the grasses nearby were different, including those in an adjacent pasture. This grass indicated the skeleton had decomposed at another location and had been moved and placed at the crash site (Figure 10.5).

Figure 10.4   (a) Plant fragments collected from a homicide victim's hair and inside her blouse were identified as lawn clippings along with petunia leaves and flowers. (b) Petunias have a very sticky glandular surface, which allows them to readily adhere to clothing and skin.

Case study 7

Circumstantial botanical evidence does not always help a case. The body of a murder victim was found in a forested area. The victim's car was found in another state. An extensive examination of the vehicle yielded a single fingerprint and numerous plant fragments.

The person to whom the fingerprint belonged admitted stealing the car, but said that he had found it elsewhere. The plant fragments in the vehicle matched the vegetation at the scene (Figure 10.6). In an attempt to have the suspect brought to the local jail for more convenient questioning, a deposition of the botanist was requested by the prosecuting attorney, who had a degree in botany. During the deposition the defense attorney, who had a degree in forestry, was able to show that the evidence not only could have come from the scene but also could have been from anywhere along a 1000-mile

Figure 10.5 This skeleton shows significant taphonomic change due to weathering from environmental exposure. Plants, seeds, and flowers associated with skeletal elements may reveal how long the remains have been in place or if they have been moved and relocated. (Photograph Courtesy of C. A. Pound Human Identification Laboratory, University of Florida) (Please refer to the colour plate section.)

track of similar woods. This circumstantial evidence was of little help in solving the crime and the suspect was convicted using other evidence, but the circumstance of the two attorneys having degrees in plant science is almost unbelievable.

Case study 8

A female victim of a sexual assault was found wandering and confused. She had no recollection of where her attack had taken place. Her hair and clothes had contained

Figure 10.6 Plant fragments recovered from a vehicle provided circumstantial evidence linking the owner of the automobile to the crime scene.

Figure 10.7   Plants fragments recovered from the hair of a sexual assault victim allowed investigators to pinpoint the site of the attack. (Courtesy of Dr. J. H. Byrd) (Please refer to the colour plate section.)

fragments of plants. The plants indicated that she had been on the ground in an area that lacked a hard surface (Figure 10.7). After identifying the plant material the law enforcement officers and a person familiar with the plants were able to scout the area where she was found and find the spot where the attack had taken place.

Case study 9

To determine if a young female victim of a sexual assault was telling the truth, her clothes were examined and her account of the attack was checked for verification. A careful examination of her clothes revealed many leaves both on the outside and the inside (Figure 10.8). The leaves were identified as being from the one species of bamboo found naturally in the area where she reported being attacked. A search of the vicinity revealed a thicket of this bamboo. The victim, taken to the thicket, recognized it. Further verification of the attack was supported by the profusion of bamboo leaves found inside her panties.

Determining time of death

Case study 1

A skeleton found in woods was beyond the normal means used by medical examiners for determining time of death. The majority of the skeleton had been moved before the botanist visited the scene. The entire area had been tramped to such an extent that the plant evidence beneath and around where the skeleton had been was no longer useful. Two bones had not been recovered at the scene, a jaw bone and a leg bone. After an intensive search, the bones were discovered, both at some distance from the original scene. The bones were left as they were found until the botanist could provide an

Figure 10.8 Fragments of bamboo leaves allowed investigators to confirm the story of a sexual assault victim and pinpoint the location of the attack. (Courtesy of Dr. J. H. Byrd)

analysis. The jaw bone was in a patch of sand and yielded no botanical evidence. The leg bone was on a fully grown leaf of a seedling of a deciduous (leaves falling in winter) tree. Under the leg bone the chlorophyll in the leaf had been shaded long enough to kill it, leaving a brown band the width of the bone. Chlorophyll in this tree takes several weeks to die. The period of weeks during which the shaded chlorophyll died was used as one time window.

The leaf of the seedling was full grown. The tip of the leaf was not distorted, which indicated that the bone had been put on the leaf after its full expansion. Residents in this rural area, aware of trees budding, gave an estimate of when spring growth started and the tree had likely budded out (Figure 10.9). The time necessary for the leaf to grow to maturity yielded a second time window.

Tooth marks on the bone indicated that it had been moved by a dog. A dog cannot rip a bone from a body until it is well deteriorated. To determine how long this deterioration would take, experienced medical examiners who worked in this area of the state were consulted. They provided an average time interval for such deterioration. The local weather service records indicated that the winter during which the decomposition occurred was unusually cold. The lower temperatures would slow the process. Extra time was added to the average, which provided the third time interval.

These three time windows, the time for chlorophyll death, the time for full leaf growth, and the time for full decomposition, were added and then subtracted from the time of discovery. A search through files for people missing near the postulated time of

Figure 10.9 The timing of budding due to spring growth was attested to by local residents to determine the length of time a body could have been located in the plant association.

death produced a match. The disappearance of the person was during the week prior to the time of death estimate.

Case study 2

The growth patterns and biology of non-woody plants can also be of value. A quite deteriorated body was found lying on the top of a bent-over plant that is a common agricultural weed. The chlorophyll was dead and the plant was beginning to rot. Time windows were again useful.

The plant was dug up and brought to a forensic anthropologist, who consulted with a forensic botanist. One method of agricultural weed control is to cover weeds with material that will completely cut off sunlight, causing the weeds to die from lack of photosynthesis. Experience has shown that a certain period of shading will kill the top of this weed. The time necessary to kill the top provided the first time window.

When the soil on the roots was removed, a shoot was found growing from the base of the plant. Research on this agricultural weed has shown shoot initiation will occur within a certain interval after the top is cut (or dies). Since the body shaded and killed the plant top, this served the same function as cutting the top. The time from top death until shoot initiation was used as the second time window.

The third time window was the estimated time for the shoot to reach the length found on the plant at the scene (Figure 10.10). These three time windows, top death, resprouting, and shoot regrowth, were added. The time of death estimated from the plant evidence corresponded to that determined from other evidence by the medical examiner.

Figure 10.10 The growth rate of this common agricultural weed (dogfennel) was utilized to determine the minimum amount of time a body would have been in place over a portion of the plant.

Case study 3

In a western area in the State of Washington, a 14-year-old girl went missing. She had last been seen with a male acquaintance 10 years her senior. An extensive search of the area surrounding the girl's semi-rural home revealed no evidence. As a result of increased scrutiny, two days after her disappearance the male acquaintance was arrested and jailed on an unrelated charge.

Two weeks following his arrest, on the advice of a psychic, the girl's body was discovered wrapped in a large roll of carpet, in a grassy field near her home. Since the area had been thoroughly searched at the time of her disappearance, a forensic botanist was called in to see if it could be determined how long the carpet had been on the site. Photographs were taken of the area beneath the carpet to document the amount of discoloration shown by the grasses. The lack of sunlight caused the discoloration called chlorosis, death of cholorophyll (Figure 10.11).

An experiment was devised to determine the amount of chlorosis that would have occurred in a given period of time. A series of 30 trash cans were filled with the correct amount of dirt to replicate the pounds per unit area of the body and placed on pieces of the same type of carpet. The trash cans and carpet pieces were then placed in the field near where the body was found. Each day one of the trash cans was removed and the area beneath it photographed. The series of 30 cans produced a clear gradient from no chlorosis to severe chlorosis, and showed that the carpet could not have been in place more than seven days (thus being placed there during a time in which the prime suspect was in jail).

Case study 4

A portion of a skeleton was found in a ditch. Plants had grown around and through the portion. Although several types of plants found at this location could be used for

Figure 10.11 Chlorosis is a term used to describe the death of cholorophyll and yellowing of green plant leaves. This process can be utilized to create an estimation of the minimum amount of time required for an object to remain in place and block sunlight from plant cells. (Please refer to the colour plate section.)

evidence of time of death, a vine was growing through the skeleton. This type of vine produces annual growth rings, which were easily interpreted (Figure 10.12). A minimum time of deposition for this partial skeleton could be accurately given the medical examiner.

Case study 5

A body was found under a common species of low-growing palm. Good color photographs were taken as the body was removed. Plants found on and in the skeleton were of little use as the area in which the body was found flooded intermittently. Soil washed over and was deposited on the skeleton on more than one occasion. Several small annual plants were growing on the uppermost deposition of soil. These plants were of no help because it was obvious from the skeletonized remains and layers of soil that the body had been at this site for several years. Since the time of soil depositions was impossible to track, the best evidence for time of death was the growth of the adjacent palms. The palms had been burned and the skeleton had not. The skeleton must have been deposited after the fire. Fire records are usually very accurate, but this fire was not recorded.

This particular palm is a common agricultural weed (Figure 10.13). Fire is a means by which the weed is controlled. Research by agriculture has found that the number of leaves produced after a fire is consistent. A certain number are produced immediately after the fire, then a stable number grows every year afterwards. The numbers of fronds

Figure 10.12 Growth rings produced in this vine can be utilized to provide time estimations for items the plant may have growth around or through.

which had grown since the fire were counted. The number of leaves produced immediately after the fire was subtracted and the remaining leaves divided by the number produced per year to indicate an average number of fronds per year. By dividing the total number of fronds by the average, a minimum time since body deposition could be determined.

Figure 10.13 Plant growth post fire damage to this palm was utilized to determine when the body was placed at the scene. Growth since the burn can be determined because the number of leaves that are produced each year is known. The provides physical evidence of a portion of the postmortem interval.

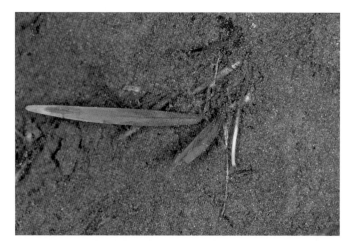

Figure 10.14 The stage of chlorosis in this grass leaf was utilized to provide a minimum time estimation for burial of the associated human remains. (Courtesy of Dr. J. H. Byrd)

Case study 6

Various pieces of plants were included when a body was buried. One of the species was identified as a common turf grass. The grass showed discoloration indicating death of chlorophyll (chlorosis) due to loss of light, which could indicate the time of the burial. The chlorosis on the grass was photographed immediately on discovery (Figure 10.14). Similar pieces of grass from this site were buried and removed at selected intervals. By matching the color of the evidence grass to the color of the experimental samples a time of burial of the grass could be estimated to within a few hours. The time to produce this particular chlorosis simply means that the grass was buried for that length of time. The body could have been placed there at that time or the burial could have been disturbed and the grass samples buried at a later time.

Index

Forensic Botany: A Practical Guide, First Edition. David W. Hall and Jason H. Byrd.
© 2012 John Wiley & Sons, Ltd. Published 2012 by John Wiley & Sons, Ltd.